NEUROINCLUSION

This book is a practical guide for anyone with managerial and leadership responsibilities. It walks you through what you need to know to develop neuroinclusive working cultures that enable, develop and get the best out of all team members.

Up to 20% of the working population has a neurodivergent difference – such as dyslexia, autism or ADHD. The diversity of thought and skills that such people bring to the workplace has been shown to increase creativity, enhance profitability and improve team morale. Managers and leaders have a key role to play in harnessing these benefits: it is up to them to create a neuroinclusive culture by setting the tone and approach.

Written by two experts in the field, both with lived experience of neurodiversity, this book is a hands-on manual for adapting your organisation to better support neurodivergent talent and developing a more inclusive and strengths-based management style. The authors provide research evidence, reflective exercises and real-life case studies to deliver an irrefutable rationale for neuroinclusivity and the techniques needed to deliver it. Managers and leaders at all levels will develop their leadership and management capabilities; and organisations will find it easier to attract and retain talent, solve problems creatively and be more resilient – critical competences in the modern working world.

Liv Blaney is an accredited executive and neurodiversity coach, and leadership training facilitator. After a successful career in London managing teams, she changed focus to supporting leaders and managers in creating inclusive workplaces through coaching and training. She has an MSc in Coaching for Behaviour Change from Henley Business School in addition to a MSc in Engineering from Cambridge University.

Sam Pringle is an accredited coach and coaching supervisor with the European Mentoring and Coaching Council and is a Presence in Action practitioner. She started a corporate career in management within the field of pharmaceuticals before developing an interest in Diversity, Equality and Inclusion with a focus on gender, parenting, caring and coaching neurodivergent employees. She has lived experience of neurodivergence. She has an MBA and an MSc in Coaching.

"A 'must have' book for every manager or people leader. This book is a brilliant combination of high-quality information, practical advice and reflective activities to develop a neuroinclusive organisation. Whether you are at the beginning of the journey to becoming neuroinclusive or a mature organisation with a well-developed approach to inclusion, I guarantee there will be something valuable in this book for you."

Louise Morpeth, *CEO, Brain in Hand*

"As someone who built a neurodivergent team at EY, this book contains all the practical wisdom I wish I'd had from day one. A game-changer for leaders serious about neuroinclusion."

Tania Martin, *Neurodiversity Consultant, PegSquared*

"This book on neurodiversity is a practical and invaluable resource for individuals, families, colleagues and organisations seeking to understand and support neurodivergent individuals. This guide challenges readers to educate themselves, reflect critically and act intentionally. Its practical approach and actionable advice make it essential for raising awareness and creating enabling environments for neurodivergent individuals."

Caroline Bixby, *Chief People Officer, Christie Group Plc*

"Liv and Sam have written the book that would have allowed me to understand not only how to be a better recruiter, leader and coach but also myself, had it been written earlier. Incredibly thorough and well-researched, it is a state-of-the-art walkthrough of best practice for inclusive leadership and universal design. This book should be on the desks of everyone who engages with others in the workplace."

Penny Terndrup, *Leadership Coach, Legal Neurodiversity Network*

"This book is a valuable resource for anyone seeking to create a more equitable and productive work environment."

Deborah Christie, *Professor of Paediatric and Adolescent Psychology*

"This brilliant book builds on fundamental management principles whilst integrating neurodiversity thinking to help managers work more effectively – and ultimately do less work. Sometimes we just need to have the right conversation, and this book equips you to do exactly that. We cannot overlook the impact leaders have on our workforce, and unless we equip them to support, manage and amplify the strengths of neurodivergent staff, we're missing our greatest opportunity."

Nathan Whitebread, *Founder, The Neurodivergent Coach*

"The book is packed full of illustrations and exercises that make it engaging, therefore increasing the chances of leading to real life and on the ground changes for a specific and historically marginalised group of employees."

James Richards, *Professor of Employee Relations,*
University of Edinburgh Business School, Herriot-Watt

NEUROINCLUSION

A Guide for Managers and Leaders

Liv Blaney and Sam Pringle

Routledge
Taylor & Francis Group
LONDON AND NEW YORK

Designed cover image: Getty Images - MarLei

First published 2026
by Routledge
4 Park Square, Milton Park, Abingdon, Oxon OX14 4RN

and by Routledge
605 Third Avenue, New York, NY 10158

Routledge is an imprint of the Taylor & Francis Group, an informa business

British Library Cataloguing-in-Publication Data
A catalogue record for this book is available from the British Library

ISBN: 978-1-041-08469-3 (hbk)
ISBN: 978-1-041-07437-3 (pbk)
ISBN: 978-1-003-64553-5 (ebk)

DOI: 10.4324/9781003645535

Typeset in Interstate
by codeMantra

CONTENTS

ACKNOWLEDGEMENTS

Thank you to all of you we have spoken to about this book and this topic. Specific acknowledgement goes to:

Those individuals whose stories are included as case studies throughout the book – thank you for being so open and giving us an insight into your world.

Pauline Macleod from the University of Edinburgh for your inquiry and curiosity.

Federico Marchiolli for your advice and sharing your practice as ND champion.

James, Kate, Gemma and the team at Heriot Watt University for your advice and encouragement as well as your on-going research in this field.

Morag Moffett Employment Lawyer and Partner at Burness & Paull LLP, for your legal advice and continued support in sharing your expertise with HR professionals and managers.

Dean Bargh from Witchwood Production House for your editing skills delivered in a detail-orientated way.

AI was not used on the sourcing or writing of content for this book.

GLOSSARY

ADHD (Attention Deficit Hyperactivity Disorder) Individuals with ADHD may have difficulties with focusing and attention, especially on tasks that the individual finds boring, repetitive or complex. They may also have difficulties with inhibiting impulses and regulating emotions.

Alexithymia A trait or characteristic which is a difficulty in recognizing, expressing and identifying emotions. For example, being able to identify if they feel happy or sad.

Asperger's syndrome This term was used as a diagnostic term for some autistic people who did not also have a diagnosis of a learning disability. Broadly, it is now agreed that what was referred to as Asperger's syndrome is part of the autism spectrum and there is no need for a separate term.[1]

Autism (Autistic Spectrum Disorder) Individuals with ASD face ongoing challenges with "social communication and interaction across multiple contexts"[2] and "restricted, repetitive patterns of behaviour, interests or activities." Autistic people may have difficulties with understanding and expressing verbal and non-verbal communication, such as tone, gesture, facial expression, and body language.

Burnout The World Health Organisation (WHO) International Disease Clarification 11[3] recognizes burnout as an occupational phenomenon. While it is not a medical or mental health condition, it has been classified as a syndrome, meaning a collection of symptoms or signs associated with a specific health-related cause.

Cognition Cognition refers to the mental action of acquiring knowledge and understanding through thought, experience, and the senses. It is the process by which we take in information and carry out an action including attention, working memory, reasoning, and problem-solving.

Comorbidity Comorbidity is the simultaneous presence of two or more medical conditions in a single individual. This can be clinical disorders or cognitive differences. Common comorbidities in the neurodiversity space include the co-occurrence of autism and ADHD, or the co-occurrence of sleep disorders and disordered eating alongside neurodivergent conditions.

Conscious/Unconscious Bias A bias is a tendency, inclination, or prejudice that results in unfair judgment, unequal treatment, or a distorted perception. Conscious bias, also known as explicit bias, refers to the attitudes or beliefs that individuals hold consciously and express openly. These biases are deliberate and intentional, and people who possess them are fully aware of their views. Unconscious bias, also known as implicit bias, refers to those attitudes or stereotypes that influence our understanding, actions, and decisions unknowingly.

Developmental coordination disorder (DCD) or Dyspraxia Individuals with DCD face challenges with physical coordination using small and large muscles. They may face challenges with balance, posture, spatial awareness, and speech as well as executive functioning. For example, handwriting, sports, and daily living tasks.

Developmental language disorder (DLD) DLD is a communication disorder that interferes with learning, understanding, and using language. It is caused by complex interactions between genes and the environment that change brain development.

Double empathy problem The double empathy problem[4] is a mutual lack of understanding of what it's like to be a neurodivergent person by a neurotypical and vice versa. This often causes breakdowns in communication.

Dyscalculia Dyscalculia impacts the ability to understand and work with numbers and mathematical concepts. Numerical processing differences can occur with language processing differences such as dyslexia.

Dysgraphia Dysgraphia affects writing, both by hand and with the keyboard. Dysgraphic people may have difficulties with handwriting, spelling, grammar, punctuation, and composition. It can also affect visual-spatial skills, memory and organisation.

Dyslexia Dyslexia is a learning difference that impacts reading, writing, spelling, memory and sometimes speaking because the brain processes language differently.

Equality Act 2010 The Equality Act 2010[5] legally protects certain categories of people from discrimination in the workplace, as well as in wider society. It applies to all employers (regardless of size) and to all workers (regardless of contract type) and job applicants. There are nine "protected characteristics" under the Act, including age, gender reassignment, race, and disability. Disability is defined as a physical or mental impairment that has a substantial and long-term adverse effect on an individual's ability to carry out normal day-to-day activities, which can include neurodivergent conditions.

Executive Function Executive functioning is a set of cognitive processes and mental skills that help an individual plan, monitor and execute their actions. This includes the ability to control attention, working memory, inhibition and problem-solving.

Hyperfocus Hyperfocus[6] is an intense state of focus, or selective attention for a sustained amount of time. This occurs only when a task is engaging for the individual and can drive performance and productivity. It may also have consequences in over-fixation to the detriment of well-being. It is a common trait of ADHD and autism.

Impulse control This refers to the ability to control your own behaviours and emotional reactions.

Intersectionality Intersectionality is the concept whereby different aspects of a person's identity – such as gender, race, class, sexuality, disability and religion – interact and intersect to create unique and complex experiences of oppression and privilege.

Mental health The World Health Organisation (WHO) defines good mental health as related to mental and psychological well-being. It is "a state of well-being in which the individual realizes his or her abilities, can cope with the normal stresses of life, work productively and fruitfully, and is able to make a contribution to his or her community".[7]

Masking / camouflaging Masking is the conscious or subconscious covering-up of neurodivergent traits to "fit in". It is particularly common in autistic and ADHD individuals, the majority of whom mask some or all the time.

Neurodivergence Cognitive function diverging from what is usually found in the population and is therefore in the minority. As everyone has neurological differences, there is naturally neurodiversity in any population. However, some individuals fall outside what is classed as the "norm" or "typical" – such as those with dyslexia, ADHD, autism – and that is classed as "neurodivergence".

Neurodiversity / neurodiverse Neurodiversity is a term that acknowledges that everyone experiences and interacts with the world around them in different ways; there is no one "right" way of thinking, learning, and behaving, and differences are not viewed as deficits.

Neuroinclusion / neuroinclusive The practice of creating environments that actively includes and accommodates individuals of all neurotypes.

Neurotypical (NT) An individual who has cognitive function that falls within what is considered "the norm," i.e. they do not have a neurodivergent condition.

Neurotype The characteristic way a person processes information, e.g. neurodivergent or neurotypical.

Obsessive Compulsive Disorder (OCD) Obsessive Compulsive Disorder is characterised by recurrent, unwanted thoughts (obsessions), and repetitive behaviours or mental acts performed to ease those thoughts (compulsions).

Pre-frontal cortex A highly evolved complex area of the brain which has key functions including executive function, decision-making, emotions, impulse control, moral judgments, attention, and focus.

Psychological safety Psychological safety[8] is a shared belief, among a group of people, that a team is safe for interpersonal risk-taking. It's the belief that we can speak up with ideas, questions, concerns and mistakes, and that we won't suffer negative social or professional consequences as a result.

Reasonable adjustments As part of the Equality Act 2010, employers must ensure that disabled employees are not at a disadvantage in the workplace due to workplace practices or physical features at work. Reasonable refers to practicality, effectiveness, cost, and disruption to others.

Rejection Sensitivity Disorder (RSD) RSD is an intense emotional reaction to the perception of being rejected by others. It can be triggered by other people's behaviours, perceived criticism, or judgment, even when their intentions were good.

Sensory difference This refers to how individuals process sensory information differently, which can affect their experiences and interactions with the world. These differences can manifest as over-sensitivity (hypersensitivity) or under-sensitivity (hyposensitivity) to sights, sounds, smells, tastes, touch, balance, or body awareness.[9]

Theory of the mind (ToM) Refers to the capacity to put yourself in someone else's shoes: to understand that others have different beliefs, behaviours, motivations, and thought processes.[10]

Tourette's / Tourette's syndrome / Tics These are disorders characterized by involuntary and repetitive movements or sounds. Tics can be motor or vocal, simple or complex, and vary in frequency, intensity, and duration; they include blinking, jumping, twirling, grunting, throat clearing, or coughing. Tics can be influenced by stress, emotions, or environmental factors. A diagnosis of Tourette's syndrome might be given if you've had several tics for over a year.

Working memory The accessible part of short-term memory which is concerned with immediate conscious processing to keep information to problem solve, remember tasks and communicate.[11]

Notes

1 National Autistic Society, "Asperger syndrome". National Autistic Society 2023. Available from: https://www.autism.org.uk/advice-and-guidance/what-is-autism/the-history-of-autism/asperger-syndrome

2 American Psychiatric Association, *Diagnostic and Statistical Manual of Mental Disorders*, 5th edn, Text Revision (DSM-5-TR). American Psychiatric Association, 2023.

3 World Health Organisation, "Burn-out an 'occupational phenomenon': International Classification of Diseases". World Health Organisation, 2019. Available from: https://www.who.int/news/item/28-05-2019-burn-out-an-occupational-phenomenon-international-classification-of-diseases

4 Milton, D., "On the ontological status of autism: the 'double empathy problem'". *Disability & Society*, 27(6) (2012), 883–87.

5 Government Equalities Office and Equality and Human Rights Commission, "Equality Act 2010". UK Government, 2010. Available from: https://www.legislation.gov.uk/ukpga/2010/15/contents

6 Ashinoff, B.K., and Abu-Akel, A., "Hyperfocus: The forgotten frontier of attention". *Psychological Research*, 85(1) (February 2012): 1–19.

7 World Health Organisation, "Health and well being". World Health Organisation, 2021. Available from: https://www.who.int/Data/Gho/Data/Major-Themes/Health-and-Well-Being

8 Edmondson, A.C., "Psychological safety and learning behaviour in work teams". *Administrative Science Quarterly*, 44(2) (1999), 350–83. Available from: https://www.researchgate.net/publication/313250589_Psychological_safety_and_learning_behavior_in_teams

9 He, J.L., Williams, Z.J., Harris, A. et al. "A working taxonomy for describing the sensory differences of autism". *Molecular Autism*, 14 (2023), 15. https://doi.org/10.1186/s13229-022-00534-1

10 Baron-Cohen, S., *Mindblindness: An Essay on Autism and Theory of Mind*. Boston, MA: MIT Press/Bradford Books, 1995.

11 Baddeley, A., "Working memory". In Alan Baddeley, Michael Eysenck, and Michael C. Anderson (eds), *Memory*, 3rd edn. Abingdon: Routledge, 2020, pp. 71–111.

1 Introduction

Why read this book?

The purpose of this book is to encourage the inclusion of neurodivergent (ND) individuals–such as those with autism, dyslexia or Attention Deficit Hyperactivity Disorder (ADHD) – in workplaces. To do this we will provide good-quality information, practical exercises and real-life case studies that can support managers and leaders in creating inclusive teams and inclusive working environments.

Our background and approach

Together, the two authors of this book have: lived experience of neurodiversity; experiences of managing and leading neurodivergent individuals, as well as coaching neurodivergent (and neurotypical) individuals; and backgrounds in challenging cultural and corporate perceptions of neurodiversity. The neurodivergent individuals whom we have managed, coached, led and learnt from have informed this book, and many of their voices are to be heard in the case studies. We have also drawn on the latest research and incorporated some reflective exercises to aid readers in thinking about how they might apply what they are learning.

Who is this book for?

- Managers of neurodivergent talent
- Co-workers of neurodivergent individuals
- Leaders of teams with neurodivergent talent
- DEI (diversity, equity and inclusion) and HR (human resources) professionals

DOI: 10.4324/9781003645535-1

Our aim is to present a simplified, accessible guide to becoming a better leader, manager or co-worker of ND individuals by reflecting on your perspective and approach around ND team members. Although the examples are from a UK perspective, with regard to legislation and culture, the overall themes and the research underpinning them are universally relevant. Given our experience and background, we have focused in the most part on office-based roles, although many of the principles and approaches used throughout may also apply to jobs that are not desk or office based.

What will this book cover?

We have chosen the topics and themes that have come up most often in our roles as managers and coaches. It is not an exhaustive list, but rather serves as a starting point for learning in this space. Those topics are:

- What "neurodiversity" is and why it has become increasingly important in the working world
- The strengths and challenges that neurodivergent individuals experience
- The benefits and challenges for managers and leaders in having an ND workforce
- How companies can adapt their systems and processes to be more neuroinclusive
- How managers and leaders can have better-quality conversations with their teams
- How to tackle biases around neurodiversity, both within yourself and among others
- Communicating effectively with ND individuals
- Creating the conditions for ND talent to thrive
- Best practices in managing performance of ND individuals
- How to manage change effectively
- Effectively utilising the strengths of ND individuals
- Considerations around health and wellbeing for neurodivergent adults

What makes this important for leaders and managers now?

In 2025, the jobs website Indeed posted a list of the skills most often sought by employers.[1] That same year, the University of London undertook a study which revealed the most common skills possessed by ND individuals.[2] You may be surprised to learn that there is a significant degree of crossover between the two. That alone makes a strong case for the benefits of hiring ND workers.

Looking further ahead, the increasing use of Artificial Intelligence (AI) is going to impact the kinds of skills that will be required in future, turning our focus towards the tasks that humans do better. A World Economic Forum report[3] (see Figure 1.1) demonstrates that a number of these future skill requirements align with well-established traits of ND individuals. For example, those with dyslexia often have a creative broader system approach than their neurotypical peers and creativity has been shown to be a trait of many ND individuals.[4]

Therefore, from an all-important skills perspective alone, we can see that the hiring and managing of ND individuals is a crucial component of a strong workforce.

Now let's look at generational shifts in perspective. Generation Z will account for 27% of the global workforce by 2030,[5] and, compared

Table 1.1 Skills sought by employers and skills offered by neurodivergent people

Skills employers most often look for[1]	*Skills ND individuals commonly bring to the workforce*[2] *and (in brackets) the proportion of ND individuals bringing that skill*
Communication skills	Verbal comprehension (44%)
Leadership skills	Authenticity (64%)
Teamwork skills	
Interpersonal skills	
Learning/adaptability	Entrepreneurialism (46%)
Self-management skills	Cognitive control (40%)
Organisational skills	Detail orientation (71%)
Computer skills	
Problem-solving skills	Creativity (78%)
Open-mindedness	Innovative thinking (75%)
Strong work ethic	Hyperfocus (ability to focus intensely for a period of time) (80%)

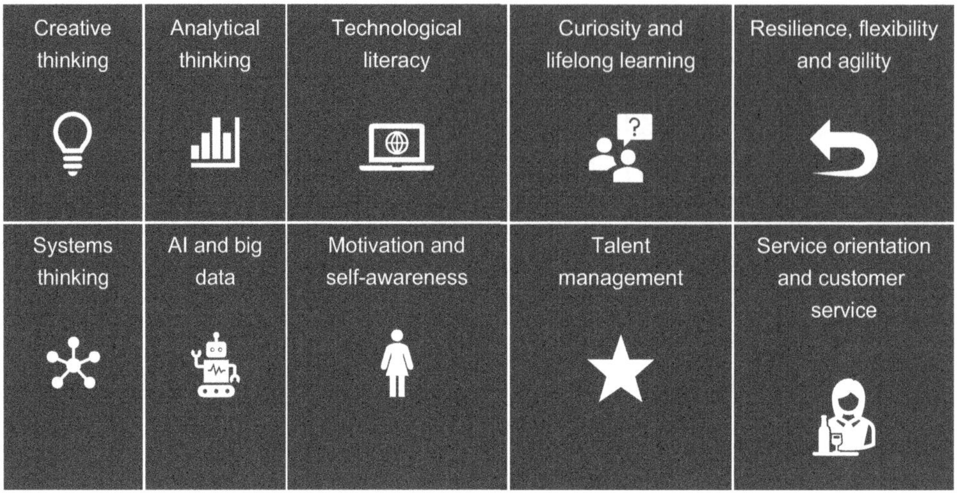

Figure 1.1 Top skills that will be required in the future

Source: World Economic Forum, The Future of Jobs Report 2023. World Economic Forum, 30 April 2023. Available at: https://www.weforum.org/publications/the-future-of-jobs-report-2023; Icons authors' own

to previous generations, they have far greater expectations of the firms that they work for on matters of inclusivity.[6] This coincides with an increasing awareness of – and diagnosis of – neurodivergence. To recruit and retain this younger talent, companies will need to be in alignment with the needs and values of their changing workforce.

And yet, a 2024 Chartered Institute of Personnel and Development (CIPD) survey in the UK found only 33% of companies included neurodiversity within their DEI strategy.[7] The same survey showed that, while 70% of companies thought that Equity Diversity and Inclusion was a critical priority for their organisations, only 60% of companies thought that neurodiversity was included in their considerations. More concerning still, a survey by the City and Guilds Foundation[8] showed that 41% of companies had no resources available around neurodiversity. Surveys have routinely shown that managers have low levels of awareness of neurodiversity, struggle with managing neurodivergent talent and find the experience to be emotionally exhausting (in part driven by the lack of support and training given to the managers).[9] Admittedly, awareness of neurodiversity has improved over the last few years,[10] but these survey results point to significant work remaining to create equitable and inclusive workplaces.

How does everyone benefit from the inclusion of ND individuals?

Western societies demonstrate a progressive trend towards inclusion of difference. This started with groups whose differences are more visible – such as gender or race – and broadened and deepened to embrace a definition of diversity which includes less visible aspects – such as neurodiversity, sexual orientation and mental health. A report in 2018 by the CIPD[11] argued that neurodiversity is the next frontier of inclusion, and that organisations need to move from a deficit-based to a strengths-based approach. Since then, there have been some reversals of attitudes around diversity and inclusion – most notably from multinationals responding to President Trump's anti-diversity policies[12] – yet many individuals and companies retain their commitment to inclusivity, even though that may require a rebranding of diversity efforts.[13]

Practical benefits of being more inclusive of ND talent includes:

- Greater **breadth of talent**. By accessing a wider and more diverse range of candidates. The up to 20% of individuals who are ND are often underemployed in comparison to their neurotypical peers;[14] for example, in the UK, just 30% of autistic adults are in employment[15] in the context of an average employment rate of 95% across the general population[16] (see Figure 1.2)
- **Resilience**. Given the pace of change and increasing uncertainty in the world, the ability to cope with stress, bounce back from

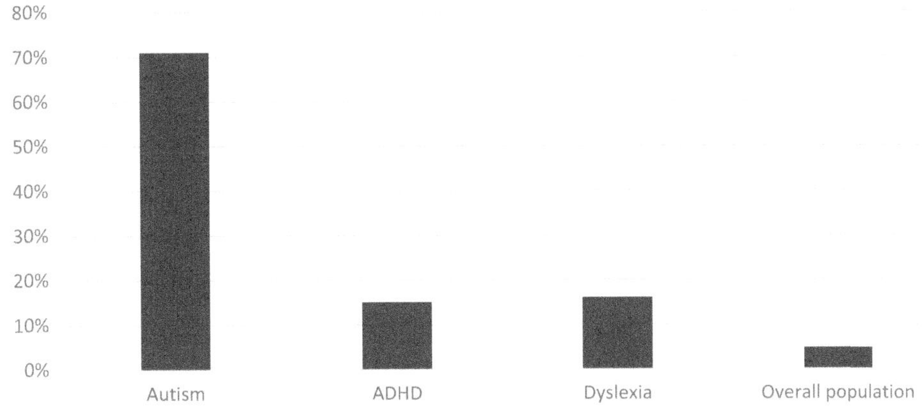

Figure 1.2 Rate of unemployment of neurodivergent individuals (UK)

adversity and learn from challenges is becoming increasingly important. And the level of resilience in any workforce has been shown to be correlated to the level of inclusion in that workforce,[17] so by creating more inclusive practices you are enabling the organisation to maintain productivity through challenging periods

- **Better retention.** Those workplaces that are more inclusive tend to have better retention.[18] One such example is the National Grid's "Disability Confident" internship scheme, which was shown to improve retention of employees across the board[19]
- **Higher productivity.** A study by Deloitte found that inclusion of ND talent can improve overall team productivity by up to 30%.[20] However, the key word there is "inclusion", as other studies have shown that, where diversity is not effectively included within a team, unconstructive conflict can result[21]
- **Greater creativity.** The diversity of thought that ND individuals bring into a team increases innovation and creative problem solving. There are numerous examples of ND thinkers (see Figure 1.3) generating new insights–from Da Vinci, Picasso and Warhol to David Neeleman (the founder of Jet Blue Airways) and Richard Branson (the founder of the Virgin brand).[22] Additionally, fostering an inclusive environment enables all team members, including neurodiverse individuals, to contribute effectively, thereby enhancing innovation.

Box 1.1: Neurodivergence and creativity

Well-known, famously creative neurodivergent individuals include:

- Agatha Christie, the famous mystery writer, who was thought to be dyslexic
- Richard Branson, the entrepreneur, who has dyslexia
- Temple Grandin, a professor of animal science, who redesigned the livestock industry through her unique thought processes, and is autistic
- Billie Eilish, the songwriter, who has Tourette's Syndrome
- Daniel Radcliffe, the actor, who has dyspraxia

- **Higher customer satisfaction and engagement.** By being inclusive of ND talent, you can better understand and serve the needs and preferences of a diverse customer base
- **Improved managerial skills.** The skills needed to successfully manage ND employees are the same skills that are effective in improving the performance of all employees. Therefore, upskilling managers and providing them with experience in managing ND employees can help improve the overall level of managerial skills

There are also significant benefits for ND individuals of being employed in an inclusive workplace. As we have seen, they may have valuable skills and qualifications but many ND people struggle to find and maintain employment. It is thought that this is caused partly by the world of work being typically constructed for the neurotypical norm, and may not be adapted to accommodate differences. Being employed provides a number of benefits (see Figure 1.3), so for ND individuals to be excluded in this way can be distressing.[23]

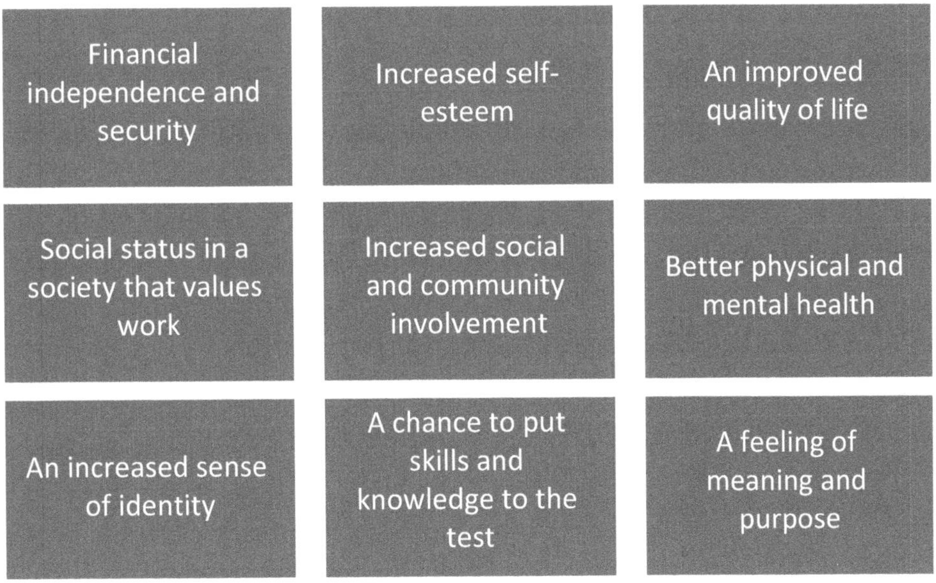

Figure 1.3 The benefits of being employed

Nonetheless, there are some companies that have been proactive in engaging with, and hiring from, the ND talent pool. Some examples are shown in Figure1.5.[24] Each of these companies created system and process changes within the organisation, and these were combined with behavioural changes by managers and leaders. We will cover both of these kinds of changes throughout this book.

Box 1.2: Companies who have capitalised on their neurodivergent talent

Company: SAP

Sector: Technology

What they did:

- The software giant launched its Autism at Work programme in 2013, which aims to recruit and retain autistic employees for various roles across the company. To date, they have hired over 240 individuals through the programme across 16 countries
- They partnered with specialist organisations, adapted their hiring and onboarding processes, set up an Employee Resources Group, provided coaching and mentoring, and created a culture of acceptance and inclusion

Some of the impact it had was:

- Increased engagement of employees through the involvement in the inclusion programme
- Improved corporate culture through the development of increased psychological safety, which is the ability for employees to speak up with ideas, questions, concerns or mistakes without the fear of punishment or humiliation
- Universal design principles were applied, which improved the workplace environment for all employees

Company: Harbour Energy

Sector: Oil and gas

What they did:

- Introduced a Neurodiversity-Focused Employee Resource Group with a senior executive sponsor, providing ND employees

with an opportunity to network, work on development and raise awareness of ND
- Created ND-friendly workspaces, including noise-cancelling earbuds and sensory rooms
- Changed recruitment practices to ensure fairness

The impact it had:

- Raised awareness internally, prompting some employees to seek diagnosis
- Provided a safe space for ND employees to share their thoughts and contribute to the strategy Provided workspaces where ND individuals can do their best work

Company: EY

Sector: Consulting

What they did:

- The professional services giant launched its Neurodiversity at Work programme in 2016, which aims to hire ND talent for roleavbs in data analytics, automation and cybersecurity. For example, they provided additional training and support and actively looked to leverage the strengths of ND employees
- They set up neurodiverse centres of excellence across a range of countries which aim to increase employment opportunities and drive a supportive and inclusive environment
- For example, EY moved from a behaviour-based interview process to a performance-based one, allowing for use of alternative methods of assessment, such as work samples, simulations or games, instead of relying on résumés and interviews. They also allow for ND individuals to have their orientations in advance, in order to allow for additional time to adapt to the new company

The impact it had:

- The benefits they have seen include increased innovation, productivity and client satisfaction, as well as a more inclusive and diverse culture

Studies have shown that, without change at managerial and leadership levels, change initiatives will not be sustained.[25] Often, change at managerial and leadership levels is encouraged through training that focuses on sharing the benefits of inclusion internally. The challenge, however, is that to change behaviour requires a change in existing assumptions, beliefs and biases – which can only be brought about effectively through the kind of reflection-based learning that is proposed in this book.[26] When your beliefs change, your behaviour changes accordingly, without having to think about it.

The effectiveness of reflective learning to achieve longer-lasting change can be explained from a neuroscientific perspective. Creating new behaviours means forming new thought patterns; to develop these patterns, new neural pathways – those that transmit messages throughout the brain – must be established. Repetition of new learning and behaviours helps to solidify these pathways, meaning that the repetition of habits is an essential component of sustained behaviour change.[27] Two key areas of the brain involved in behaviour change are the prefrontal cortex and the hippocampus.

The prefrontal cortex manages higher-order functions such as planning, decision-making and self-regulation. Strengthening connections with, and within, this brain region through repetition allows new information to be integrated into these processes. Additionally, reflecting on experiences can promote the formation of long-term memories in the hippocampus, which serves as the brain's memory hub. This facilitates easier retrieval of information and enhances future learning, thereby increasing the likelihood of sustained behaviour change.

Furthermore, your brain benefits from having a clear motivation for change. When you have defined a clear goal with a personal meaning behind it, your brain releases dopamine, a neurotransmitter (brain chemical) that helps you to stay focused and persist through the discomfort that inevitably comes with change. Therefore, it is helpful to start any behavioural change process – whether visiting the gym every day, practising some new influencing skills or becoming a more neuroinclusive leader – by thinking about what the motivation is for change, and what the outcome would look like once you've achieved that change.

Reflecting on your motivations

Throughout this book there will be self-reflection exercises like the one below. We invite you to fill them in as you progress through the book: they will help you assess your own levels of awareness and encourage you to think about past experiences from different perspectives.

To begin with, let's consider your motivation for reading this book. Try answering the following questions:

What made you interested in this topic and persuaded you to pick up this book?

....

....

....

....

For you, what would be a good outcome of reading this book? What might be different for you after you have read this book?

....

....

....

....

What would that provide to your team and other co-workers?

....

....

....

....

What impact might that have on your organisation's broader approach to neuroinclusion?

....

....

....

....

Reading this book may have a personal impact on you and how you interpret your own differences, as well as of the differences of those around you, in both your personal and professional life. We return to this point at the end of this book (Chapter 11), where we invite you to consider what actions you want to take in response to what you have learnt about yourself.

Key points from Chapter 1

1. There are significant benefits in creating a more inclusive approach to ND talent, from societal, organisational, team and individual perspectives.
2. Managers and leaders have an important role to play in modelling an inclusive culture through their behaviours.
3. To create a more inclusive culture, we need to change perspectives, by becoming aware of our biases, assumptions and automatic ways of doing things.
4. An effective way of shifting behaviour is through self-reflection: the approach used in the exercises in this book.
5. This book also provides knowledge, concepts and models that can help facilitate and support change.

Notes

1 Indeed Editorial Team "Top 11 skills employers look for in job candidates". Indeed, 29 January 2025. Available: at https://www.indeed.com/career-advice/resumes-cover-letters/skills-employers-look-for

2 McDowall, A., Doyle, N., and Srinivasan, A., *Neurodiversity in Business: Research Report 2024*. Birkbeck: University of London, 2024. Available at:

https://www.neurodiversityinbusiness.org/research/nib-and-university-of-birkbeck-research-report-2024

3 World Economic Forum, *The Future of Jobs Report 2023*. 30 April 2023. Available at: https://www.weforum.org/publications/the-future-of-jobs-report-2023

4 McDowall, A., Doyle, N., and Srinivasan, A., *Neurodiversity in Business: Research Report 2024*. Birkbeck: University of London, 2024. Available at: https://www.neurodiversityinbusiness.org/research/nib-and-university-of-birkbeck-research-report-2024

5 Koop, A., "Chart: How Gen Z employment levels compare in OECD countries". World Economic Forum, 26 March 2021. Available at: https://www.weforum.org/stories/2021/03/gen-z-unemployment-chart-global-comparisons

6 Fruitful Insights, *Gen Z–Shaping the Future of UK Workplaces: Business Insights Report 2023*. Available at: https://www.legalandgeneral.com/landg-assets/adviser/files/workplace-benefits/protection/added-value-benefits/gen-zshaping-the-future-of-uk-workplaces.pdf

7 CIPD (Chartered Institute of Personnel and Development), *Neuroinclusion at Work*. February 2024. Available at: https://www.cipd.org/globalassets/media/knowledge/knowledge-hub/reports/2024-pdfs/2024-neuroinclusion-at-work-report-8545.pdf

8 City and Guilds, *Neurodiversity Index 2025*. London: City and Guilds, 2025. Available at: https://cityandguildsfoundation.org/what-we-offer/campaigning/neurodiversity-index

9 Richards, J., Sang, K., Marks, A., and Gill, S., "'I've found it extremely draining': Emotional labour and the lived experience of line managing neurodiversity". *Personnel Review* 48(7) (2019), 1903–23. https://doi.org/10.1108/PR-08-2018-0289

10 Google Trends shows a 250% increase in the number of searches for "neurodiversity" in the UK from January 2022 to January 2025. Source: Google Trends, 2025. Available at: https://trends.google.com/trends

11 CIPD (Chartered Institute of Personnel and Development), *Neurodiversity at Work*. February 2018. Available at: https://www.cipd.org/globalassets/media/knowledge/knowledge-hub/guides/neurodiversity-at-work_2018_tcm18-37852.pdf

12 Murray, C., and Bohannon, M., "IBM reportedly walks back diversity policies, citing 'inherent tensions': Here are all the companies rolling back DEI programs". *Forbes*, 11 April 2025. Available at: https://www.forbes.com/sites/conormurray/2025/04/11/ibm-reportedly-walks-back-diversity-policies-citing-inherent-tensions-here-are-all-the-companies-rolling-back-dei-programs

13 *Newsweek*, "List of US companies that are continuing with DEI policies". *Newsweek*, 18 February 2025. Available at: https://www.newsweek.com/dei-companies-program-list-policies-donald-trump-2032960 Zheng, L., "What comes after DEI?" *Harvard Business Review*, January 2025. Available at: https://hbr.org/2025/01/what-comes-after-dei

14 Commonly cited statistics for neurodiversity are that between 15% and 20% of the population is neurodivergent (e.g. CBBC, "What is neurodiversity, how do we celebrate it, and what does neurodivergent mean?". BBC, 10 June 2024. Available

at: https://www.bbc.co.uk/newsround/64951690) The number is difficult to estimate for a few reasons, not least historic under-diagnosis, the complexity of diagnostic criteria, stigma and self-diagnosis reducing the numbers coming forwards for diagnosis, and cultural/social influences. Some of these percentages are an under-representation (e.g. Westgarth, D., "Why a neurodiversity diagnosis matters". *BDJ in Practice*, 37 [2024]: 284–85. https://doi.org/10.1038/s41404-024-2813-8). Underemployment references include the Department of Work and Pensions (UK), "Employment prospects for neurodiverse people set to be boosted with launch of new expert panel". Press release, 30 January 2025. Available at: https://www.gov.uk/government/news/employment-prospects-for-neurodiverse-people-set-to-be-boosted-with-launch-of-new-expert-panel

15 Department of Work and Pensions, UK, *The Buckland Review of Autism Employment: Report and Recommendations*. UK Government, 2024. Available at: https://www.gov.uk/government/publications/the-buckland-review-of-autism-employment-report-and-recommendations

16 Office for National Statistics, "Employment in the UK: December 2024". Office for National Statistics, 2024. Available at: https://backup.ons.gov.uk/wp-content/uploads/sites/3/2024/12/Employment-in-the-UK-December-2024.pdf

17 Hundschell, A., Backmann, J., Tian, A.W. and Hoegl, M., "Leader inclusiveness and team resilience capacity in multinational teams: The role of organizational diversity climate". *Journal of Organisational Behaviour*, 46(3) (2024). https://doi.org/10.1002/job.2829. Gong, L., Zhang, S. and Liu, Z., "The impact of inclusive leadership on task performance: A moderated mediation model of resilience capacity and work meaningfulness". *Baltic Journal of Management*, 19(1) (August 2023). https://doi.org/10.1108/BJM-01-2023-0029. Duchek, S., Raetze, S., and Scheuch, I., "The role of diversity in organisational resilience: A theoretical framework". *Business Research*, 13 (2020), 387–423. https://doi.org/10.1007/s40685-019-0084-8

18 Ali, M., Grabarski, M.K. and Baker, M., "The influence of neurodiversity management on affective commitment and turnover intention: The role of neurodiversity awareness". *Employee Relations: The International Journal* 46(3) (2024), 657–74. https://doi.org/10.1108/ER-09-2023-0476

19 National Grid, "Building work skills for young people with additional needs". National Grid, 3 December 2020. Available at: https://www.nationalgrid.com/stories/community-spirit-stories/building-work-skills-young-people-additional-needs

20 Mahto, M., Hogan, S.K., and Snideman, B., "A rising tide lifts all boats: Creating a better work environment for all by embracing neurodiversity". Deloitte, 18 January 2022. Available at: https://www2.deloitte.com/us/en/insights/topics/talent/neurodiversity-in-the-workplace.html

21 Jehn, K.A., "Diversity, conflict, and team performances: Summary of program of research". *Performance Improvement Quarterly*, 12(1) (2008), 6–19. https://doi.org/10.1111/j.1937-8327.1999.tb00112.x

22 Siegel, L.S., "Agatha Christie's learning disability". *Canadian Psychology*, 29(2) (1988), 213–16. https://doi.org/10.1037/h0084531. MacDonald-Smith, F.,

"Daniel Radcliffe: Breaking dyspraxia's spell". *The Telegraph*, 15 September 2008. Available at: https://www.telegraph.co.uk/news/health/3356396/Daniel-Radcliffe-breaking-dyspraxias-spell.html Pandey, M., "Billie Eilish: Living with Tourette's is 'very exhausting'". BBC, 24 May 2022. Available at: https://www.bbc.co.uk/news/newsbeat-61565717. Branson, R., "Launching DyslexicU: The world's first 'University of Dyslexic Thinking'". Virgin, 2024. Available at: https://www.virgin.com/branson-family/richard-branson-blog/launching-dyslexicu-the-worlds-first-university-of-dyslexic-thinking Grandin, T., "When great minds don't think alike". *Harvard Business Review*, May 2023. Available at: https://hbr.org/2023/05/when-great-minds-dont-think-alike

23 Hagler, M., Hamby, S., Grych, J., et al., "Working for well-being: Uncovering the protective benefits of work through mixed methods analysis". *Journal of Happiness Studies* 17(4) (2016), 1493–510. https://doi.org/10.1007/s10902-015-9654-4. Public Health England, "Health matters: Health and work". UK Government, 2019. Available at: https://www.gov.uk/government/publications/health-matters-health-and-work/health-matters-health-and-work. Schob, R., "Unemployment and identity". *CESifo Economic Studies* 59(1) (March 2013), 149–80. https://doi.org/10.1093/cesifo/ifs040. Ibañez, L.M. and Lopez, S.H., "'Coming back to who I am': Unemployment, identity, and social support". *Race, Identity and Work* (*Research in the Sociology of Work* 32). Leeds, UK: Emerald Publishing, 7–33. https://doi.org/10.1108/S0277-283320180000032004.

24 Loucks, S., "SAP's autism at work program celebrates 10 years of success". SAP, 18 July 2023. Available at: https://community.sap.com/t5/human-capital-management-blogs-by-sap/sap-s-autism-at-work-program-celebrates-10-years-of-success/ba-p/13562404?form=MG0AV3. Harbour Energy, "Diversity, equity and inclusion". Harbour Energy, 2025. Availableat:https://www.harbourenergy.com/safety-esg/social/diversity-equity-and-inclusion. City and Guilds, *Neurodiversity Index 2025*. London: City and Guilds, 2025. Available at: https://cityandguildsfoundation.org/what-we-offer/campaigning/neurodiversity-index. Raz, M., "How top companies are leading the charge for a neuroinclusive workplace". *Forbes*, 11 October 2024. Available at: https://www.forbes.com/councils/forbescoachescouncil/2024/10/11/how-top-companies-are-leading-the-charge-for-a-neuroinclusive-workplace. EY UK, "EY launches first neurodiverse centre of excellence in the UK to boost client innovation". Press release, 6 July 2021. Available at: https://www.ey.com/en_uk/newsroom/2021/07/ey-launches-first-neuro-diverse-centre-of-excellence-in-the-uk-to-boost-client-innovation

25 Coffman, J., Bax, B., Noether, A., and Blair, B., "The fabric of belonging: How to weave an inclusive culture". Bain and Company, 2022. Available at: https://www.bain.com/insights/the-fabric-of-belonging-how-to-weave-an-inclusive-culture. Enei, *Inclusive Leadership... Driving Performance through Diversity!* 2016. Available at: https://www.cipd.org/globalassets/media/knowledge/knowledge-hub/reports/inclusive-leadership_2016-driving-performance-through-diversity_tcm18-8811.pdf Jefferson, P. "Aligning DEI initiatives with organisational strategy: A case

study of DEI professionals in global technology firms". University of Arizona Global Campus ProQuest Dissertation and Theses, 2025.
26 Brooks, D., *How to Know a Person*. London: Penguin, 2023.
27 Brann, A., *Neuroscience for Coaches: How to Use the Latest Insights for the Benefit of Your Clients*. London: Kogan-Page, 2014.

2 What is neurodiversity?

Imagine a paperclip. Now, let's say you spend five minutes considering all the potential uses for that paperclip, being as creative as possible. The uses that you come up with are likely to be different (at least in part) to those that your colleagues would come up with - because *everyone thinks differently*. And everyone thinks differently because our brains are all wired differently. This is the basis of the idea of neurodiversity: everyone's brains are naturally different, giving rise to different strengths and challenges in different individuals. However, while all brains work differently, and they all have different strengths and challenges, there are brains that work significantly differently to what most people would consider to be "typical". The term applied to those whose brains work differently to what most people would think of as the "typical" - such as those with autism, dyslexia or attention-deficit hyperactivity disorder (ADHD) - is "neurodivergent" (ND).

Box 2.1: The definition of neurodiversity

Neurodiversity is a term that acknowledges that everyone experiences and interacts with the world around them in different ways; there is no one "right" way of thinking, learning and behaving, and differences are not viewed as deficits.

The past: medicalising neurodiversity

There are a number of different models to understand neurological difference and formulate ways of addressing it, the principal two

DOI: 10.4324/9781003645535-2

being the Medical Model and the Social Model, which contrast with one another on account of their basic assumptions.

The Medical Model

The Medical Model relies on the assumption that there is a "norm". Behaviours or characteristics that fall outside of that norm are therefore seen as impairments to be healed or cured.[1] While this approach might be applicable to physiological issues such as a broken leg, or diabetes, it is problematic when it comes to the diagnosis and treatment of ND conditions, such as dyslexia, ADHD and autism.

Under the Medical Model, these conditions are to be "fixed" to render the subject "normally functioning". This is something that many within the disability community find frustrating, given that they may not be willing, able or have a desire to become "normally functioning". The medicalisation of ND implies and reinforces the idea that people who are diagnosed with neurodivergent conditions are less worthy and/or capable than others. These are the attitudes that lead ND individuals to "mask" their differences in an attempt to fit in with the majority and be considered equal;[2] see Box 2.2 for more information.

Box 2.2: Masking

Masking is the conscious or subconscious covering-up of neurodivergent traits in order to "fit in". It is sometimes also known as "camouflaging". It is particularly common in autistic and ADHD individuals, the majority of whom mask some or all of the time. Behaviours that are adopted in masking include:

- Mirroring others' facial expressions
- Forcing yourself to make eye contact, even if its uncomfortable
- "Scripting" conversations ahead of time

These are often done consciously and reactively with the aim of "fitting in". But masking typically leads to feelings of discomfort and anxiety. Over time, this pretence of fitting in can be exhausting, and can result in isolation, poor mental health, loss of identity and delayed diagnosis. It is thought to be a contributing factor in the relatively high incidence of death by suicide in people with autism. It is worth noting that neurotypical individuals also mask, but seemingly less often, and with reduced personal impact.

Source: Evans, J., Krumrei-Mancuso, E.J. and Rouse, S.V., "What you are hiding could be hurting you: Autistic masking in relation to mental health, interpersonal trauma, authenticity, and self-esteem". *Autism in Adulthood*, 6(2) (2022). https://doi.org/10.1089/aut.2022.0115. Kelly, C.A., Kelly, C. and Taylor, R., "Review of the psychosocial consequences of Attention Deficit Hyperactivity Disorder (ADHD) in females". *European Journal of Medical and Health Sciences*, 6(1) (2024), 10–20. https://doi.org/10.24018/ejmed.2024.6.1.2033

The Social Model

Under the Social Model, society is seen as the cause of the barriers and challenges that prevent ND individuals from fully participating in work and society. These barriers tend to fall under three categories:[3]

Box 2.3: The Social Model of Disability

This model says that people are disabled by barriers in society, not by their impairment or difference. Barriers can be environmental, social or institutional in nature. Removing these barriers creates equality and offers disabled people more independence, choice and control.

1. **Environmental**. This includes inaccessible infrastructure and poor service provision. For instance, websites might lack visual elements or have unclear instructions, making them hard to navigate.

Similarly, educational or work environments might be constantly noisy with no quiet areas, causing sensory overwhelm for certain autistic or ADHD individuals.

2. **Societal**. Perceptions of neurodiversity have been shaped by the autism stereotype of the geeky, clumsy and socially awkward schoolboy, or the ADHD stereotype of the hyperactive, distracted schoolboy. Over time, society's understanding of neurodivergence and how it shows up has evolved, but the stereotypes, and the biases, haven't evolved along with it. For example, a 2020 Institute of Leadership report[4] found that half of all leaders and managers would not employ someone who had one or more neurodivergent conditions, with many providing a reason that was based around assumptions of incapability.

   The media plays a big part in shaping attitudes. And its current narrative around ND is a mixed one. For example, in the UK *Times* newspaper in March 2024, columnist Matthew Parris declared that he doesn't believe in ADHD "except as a catch-all for a whole gaggle of unrelated ways people behave".[5] Other journalists take a more inclusive approach: articles in the UK *Guardian* have flagged the multiple challenges for ND individuals in getting support and diagnosis.[6] It makes for a confusing picture for those who have not fully investigated the neurodiversity space.

3. **Institutional**. This refers to government and organisational policies and processes that mandate inclusive practice. The kinds of areas they might address are:
   a. Interview processes that are based on your ability to respond quickly to questions, yet such an ability not being a requirement for the role. Quick responses in high-pressure situations can be challenging for those with conditions like autism, who may absorb and process information at a different speed to others
   b. Educational settings that do not account for a diversity of learning styles. They might rely predominantly on reading and writing skills to support learning rather than using more practical or physical exercises. This can be challenging for

individuals with conditions such as dyslexia, and is a likely reason why a significant proportion of school non-attenders are neurodivergent[7]

Critics of the Social Model of Disability have said that, even if you took away all the barriers to inclusion, individuals would still face significant challenges.[8] For example, an individual with ADHD who struggles with time management will still face these challenges even if they have access to time reminder and prioritisation apps: i.e. the issues are not simply created by the wider world.

At of the time of writing, there is limited agreement on what model to use, and alternatives have been proposed. One is the Biopsychosocial Model (BPSM), which includes aspects of both the Medical (biological) Model and the Social Model of Disability while also incorporating individual variations influenced by beliefs and attitudes and how they impact behaviour at work (psychological). It allows for greater complexity such that some individuals may benefit from medical intervention (e.g. drugs, therapy) and systemic accommodations while not pathologising the conditions.

The key point is that the concept of neurodiversity acknowledges that there is a natural variation in neurological processing. And it further implies that, in some cases, whether driven by society or by behavioural capabilities, adaptations or additional support may be required.

Myth buster

This chapter will follow with details about each neurodiversity diagnosis. But before that we invite you to reflect on your current level of understanding and assumptions about this topic. Rate yourself against each of the statements in Table 2.1, from 1 (strongly disagree) to 5 (strongly agree). Bear in mind that there are both facts and fictions in this table. As we go through the book, we will learn to distinguish one from the other.

Table 2.1 Myth buster

Statement	*Strongly agree*	*Slightly agree*	*Not sure/ neutral*	*Slightly disagree*	*Strongly disagree*
Neurodiversity					
Parents should keep their child's neurodivergence a secret					
If you discipline your child correctly, they will learn to manage their neurodivergence properly					
Neurodivergent individuals can't form healthy relationships					
Neurodivergent individuals can have successful careers					
The traits that neurodivergent individuals have vary through life					
We are all neurodivergent in some way, as we are all on a spectrum					
Managing neurodivergent individuals takes more time than managing neurotypical individuals					
Autism					
Autistic individuals can be extroverts					
Autistic people cannot handle stress					
Autism can be cured					
Autistic individuals find it difficult to empathise with others					
Autistic people like social interaction					
Autism is a result of poor parenting					
Autistic people are all the same: smart but socially awkward					

(Continued)

Table 2.1 (Continued)

Statement	*Strongly agree*	*Slightly agree*	*Not sure/ neutral*	*Slightly disagree*	*Strongly disagree*
We are all on the "autism spectrum" in some way, as we all have some autistic traits					
Autistic individuals have higher-than-average IQ					
More men than women are autistic					
Autistic individuals are best placed in analytical careers like technology					
Autism is diagnosed based in part on an individual experiencing difficulties with social interactions					
Medication can help those with autism live a normal life					
ADHD					
ADHD isn't a real medical condition					
People with ADHD can solve the challenges they face if they try hard, although this can be exhausting					
ADHD is over-diagnosed					
ADHD medicines like Ritalin can benefit everyone					
Medication is the only treatment for those with ADHD					
People with ADHD can't ever focus					
Kids with ADHD are hyperactive					
Women and men are just as likely to have ADHD as each other					
People with ADHD are best off being self-employed; they can never fit within organisations					
You can grow out of ADHD					

(Continued)

Table 2.1 (Continued)

Statement	*Strongly agree*	*Slightly agree*	*Not sure/ neutral*	*Slightly disagree*	*Strongly disagree*
Obsessive Compulsive Disorder (OCD)					
OCD is just about being clean or organised					
OCD is caused by a troubled childhood					
Diagnosable OCD is rare					
OCD isn't treatable					
People with OCD can stop their compulsions if they just try hard enough					
People with OCD are always a bit neurotic					
Dyslexia					
Writing letters backwards is the main sign of dyslexia					
Dyslexia is caused by poor teaching					
Dyslexia only affects reading					
Dyslexia can be outgrown					
Dyslexic people have lower-than-average IQ					
Dyslexia is a vision problem					
Dyspraxia					
Dyspraxia is being clumsy and doesn't need adaptations at work					
Dyspraxia is only about coordination					
People grow out of dyspraxia in their teenage years					
Dyscalculia					
Dyscalculia is "maths dyslexia"					
Dyscalculia is an extreme form of having anxiety about maths					
You can grow out of, or be trained out of, dyscalculia					
Dyscalculia is always diagnosed alongside dyslexia					

(Continued)

Table 2.1 (Continued)

Statement	*Strongly agree*	*Slightly agree*	*Not sure/ neutral*	*Slightly disagree*	*Strongly disagree*
Dysgraphia					
Dysgraphia means you struggle with diagrams					
You can grow out of dysgraphia					
Tourette's/tics					
People with Tourette's are always shouting words at inappropriate times					
People with Tourette's could control themselves if they try hard					
Everyone with Tourette's blurts out obscenities					
People can grow out of Tourette's					
Tourette's is a mental illness					
There are just two types of tic: motor tics and vocal tics					
People can grow out of tics in their teenage years					
Tics are contagious: i.e. if I start tic-ing, then you - if you are known to have tics - are more likely to start tic-ing					

The current state (in the UK)

Over the last few years, there has been a significant increase in focus on ND conditions.[9] As a result:

- Schools have become more aware of the different representations of ND conditions, particularly ADHD and autism. This is thought to have resulted in an increase in children being diagnosed, although the numbers tracking this are inconsistent. Estimates suggest that a significant proportion of school non-attenders are neurodivergent,[10] with non-attendance driven at least in part by a school system that fails to fully accommodate their needs (e.g. classes that demand constant focus, being seated, for long periods - a potential challenge for those with ADHD)[11]

- There has been an increase in adult diagnoses.[12] This is in part driven by parents seeking help, having perceived a link with their own child's neurodivergent diagnosis. This is particularly true for women, who have historically been overlooked for diagnosis at school age.[13] As a result, there are now long waits in the UK for diagnosis. Parents in workplace leadership positions are now more frequently looking to their organisations to provide support for them, both as parents of ND children and as ND adults themselves
- There has been significant media and social media coverage of neurodiversity,[14] which includes narratives about the rising rates of diagnoses and the failure of the healthcare system. Celebrities have been talking about their neurodivergence and how it has helped them in their careers.[15] However, such "isn't my neurodivergence great?" narratives aren't always well received, for two reasons. Firstly, celebrities' access to resources (e.g. diagnosis and talking therapy) doesn't reflect the experiences of most of the population, where there are sometimes long waits for support provision. Secondly, celebrities have in some cases reframed their neurodivergence as a gift in their chosen career - high-functioning "superhuman" or "genius" - not a deficit which they have overcome. This "superpower" narrative can create unrealistic expectations of ND individuals at work, generally

A note on language

The language used to describe neurodiversity and neurodivergence is a sensitive and controversial issue (see Box 2.4). It is also worth noting that there is thought to be a higher proportion of gender non-binary individuals within the neurodivergent population.[16] Therefore, in addition to clarifying any preferred language around neurodivergent conditions, it can also be worth being mindful of any preferred pronouns.

Box 2.4: Terminology

Neurodivergent - having cognitive functioning different from that which is viewed as "the norm" or "typical". Neurodivergent brains process information differently to most of the population.

This can be genetically driven neurodivergence, such as ADHD or autism, as well as neurodivergence caused by brain changes (for example, head injury).

Neurodivergence - cognitive function diverging from what is usually found in the population and therefore in the minority. As everyone is in some way neurologically different to everyone else, there is naturally neurodiversity in any population. However, some individuals - such as those with dyslexia, ADHD, autism - fall outside what is classed as "the norm" or "typical", and that is described as "neurodivergence".

Neurotypical - an individual who has cognitive function that falls within what is considered "the norm", i.e. they do not have a neurodivergent condition.

Person-first language - each person will have their own preferences around how they describe themselves. One option is "person-first" language such as "a person with ADHD", which emphasises the person before the condition, to try to avoid being defined primarily by that trait or condition. Others see their difference as an integral part of who they are, and might prefer a phrase such as "an ADHD person". In this book, we use both versions, reflecting the variety of preferences in this space.

Executive functioning - a set of cognitive processes and mental skills that help an individual plan, monitor and execute their actions. This includes the ability to control attention, working memory, inhibition and problem-solving. These processes are often thought to originate in the brain's prefrontal cortex.

Neurotype - the characteristic way a person processes information, e.g. neurodivergent or neurotypical.

Neuroinclusion - the practice of creating environments that actively include and accommodate individuals of all neurotypes.

Ableism - discrimination, prejudice or social bias against people with disabilities, typically rooted in the belief that able-bodied people are superior. This results in attitudes, behaviours, policies and practices that marginalise or exclude people with physical, sensory, intellectual or mental health disabilities.

Common neurodivergent diagnoses

The main neurodivergent diagnoses are dyslexia, dyspraxia, dyscalculia, dysgraphia, obsessive-compulsive disorder (OCD), ADHD, autism and Tourette's/tics. These diagnoses are not mutually exclusive, and it is common for people to have traits of more than one condition. For example, 20–65% of neurodivergent people have both autism and ADHD (AuADHD).[17] It is for this reason that throughout this book we will tend to refer to "neurodivergent" individuals or specific neurodivergent traits, rather than specific medical conditions or diagnosis. There is some debate about the scope of what falls under the term "neurodivergent" and "neurodiversity"; what follows represents the most commonly held view that we have come across in studies. It is

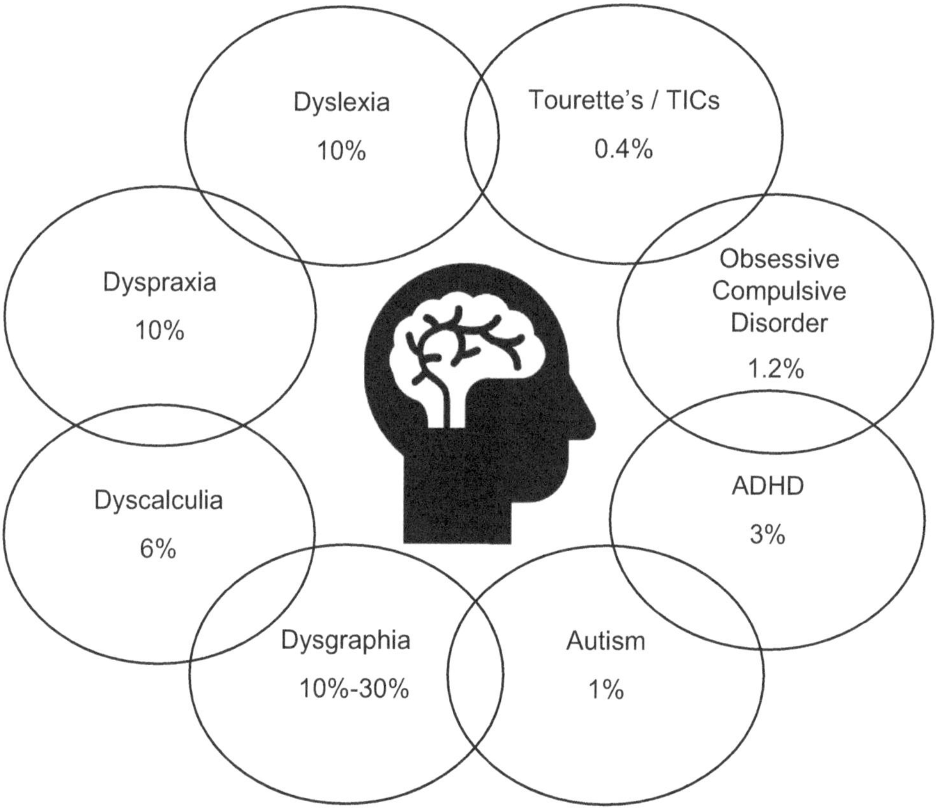

Figure 2.1 Neurodivergent conditions and their prevalence in the population

All of the numbers cited are estimates, with significant variations in the prevalence across different sources, which are as follows: On dyscalculia[18]; on dyslexia[19]; on dysgraphia[20]; on dyspraxia[21]; on ADHD[22]; on autism[23]; on OCD[24], and on Tourette's/tics[25]

also worth noting that there is some debate about the way in which terms are expressed. For example, some individuals prefer usage of Attention Deficit Hyperactivity Condition (ADHC) rather than ADHD, reflecting a shift towards a more societal model of disability. What we reflect here is what is used most commonly at time of writing.

Here are some brief descriptions of each diagnosis and some of the strengths and challenges they may confer in the workplace for some people. It is important to note that these are generalisations: each person's strengths and challenges are unique to the individual and both challenges and strengths are often on a continuum. For example, some individuals are not able to live and work independently.

Dyslexia

Dyslexia is a learning difference impacting reading, writing, spelling, memory and sometimes speaking. It is most commonly diagnosed at school, when learning differences, in comparison to a student's peers, are observed by teachers.

- Strengths in the workplace may include:
  - Creativity
  - An ability to see the bigger picture
  - Problem-solving capabilities
  - Resourcefulness
  - Pattern identification
- Challenges in the workplace may include:
  - A slow writing or reading speed
  - Writing or reading inaccurately
  - Confusing sounds and letters (in some cases)
  - Not being able to engage with large amounts of text
  - Appearing to be disorganised
  - Poor short-term memory, e.g. remembering instructions

Dyspraxia

Dyspraxia (also known as developmental coordination disorder) affects movement, coordination and sensory processing. Dyspraxic people may have difficulties with the movements of both small and large muscles that are used to, for example, button a shirt, write, run or climb. Additionally, they may face challenges with balance, posture, spatial awareness and speech. Dyspraxia can also affect planning, memory, organisation and executive functioning.

- Strengths in the workplace may include:
  - Big-picture thinking
  - An ability to spot patterns
  - Reasoning/critical thinking
- Challenges in the workplace may include:
  - Tasks requiring coordination, e.g. using a computer mouse
  - Not accurately reading social cues/body language
  - Not being aware of personal space
  - Becoming overwhelmed with sensory inputs
  - Poor short-term memory, e.g. remembering instructions
  - Finding their way around physical environments, i.e. following maps
  - Managing emotions

Dysgraphia

Dysgraphia is a learning difference that affects writing, both by hand and by keyboard. Dysgraphic people may have difficulties with handwriting, spelling, grammar, punctuation and composition. It can also affect visual-spatial skills, memory and organisation.

- Strengths in the workplace may include:
  - Listening skills
  - Oral storytelling
  - Comfort with expressing themselves verbally
- Challenges in the workplace may include:
  - Untidy, inconsistent or illegible handwriting
  - Inaccurate spelling and punctuation
  - Difficulty structuring sentences and therefore expressing themselves in writing

Dyscalculia

Dyscalculia impacts an individual's ability to understand and work with numbers and mathematical concepts. Often, it is thought to be the "numbers-based" version of dyslexia, as it concerns numerical challenges rather than language-based ones.

- Strengths in the workplace may include:
  - Creativity
  - Strategic thinking
  - Intuitive reasoning
- Challenges in the workplace may include:
  - Numerical tasks, e.g. calculations, budgeting, data analysis
  - Time management
  - Misreading mathematical symbols
  - Finding shape, direction and orientation difficult

OCD

Obsessive-compulsive disorder (OCD) involves reoccurring, intrusive thoughts and specific behaviours. Obsessive thoughts are often unwanted and can cause distress and anxiety. Compulsive behaviours are repetitive and ritualistic actions that are performed to reduce the anxiety caused by the obsessive thoughts.

- Strengths in the workplace may include:
  - Attention to detail
  - A strong work ethic
  - Perfectionism
  - Problem-solving skills
  - Reliability
- Challenges in the workplace may include:
  - Anxiety and stress
  - Compulsions interfering with daily activities, leading to social isolation
  - Difficulty with decision-making

There is some debate in the neurodiversity community about whether OCD should fall under the umbrella of "neurodiversity" or be seen solely as a mental health condition. In the reference work on mental health and brain-related conditions and disorders (known as the DSM-5),[26] OCD is part of a broader set of disorders called Obsessive Compulsive Spectrum Disorders, which includes, among others, hoarding disorder (a resistance to discarding your belongings) and skin-picking disorder.

ADHD

Attention-deficit hyperactivity disorder (ADHD) affects attention, impulsivity and hyperactivity. Individuals with ADHD may have difficulties with focusing and attention, especially on tasks that the individual finds boring, repetitive or complex. They may also have difficulties with inhibiting impulses and regulating emotions.

- Strengths in the workplace may include:
  - Quick thinking and insightfulness
  - Creative problem-solving
  - Enthusiasm and energy
  - Hyperfocus (the ability to deeply focus on a particular task, often linked to high productivity)
- Challenges in the workplace may include:
  - Multitasking or switching between tasks
  - Lack of attention to some details and over-attention to others
  - Disorganisation, poor planning and time management
  - Memory and forgetfulness
  - Distractibility
  - Task paralysis, e.g. getting started on tasks, finishing tasks

Box 2.5: ADD and ADHD

Historically, ADHD-ers have been perceived to have a consistent lack of focus, which was reflected in the term that preceded it: attention deficit disorder (ADD). However, the reality is that many ADHD-ers can experience both distractibility and "hyperfocus", the latter being an ability to intensely focus on a task for a prolonged period of time. This means that, when interested in a topic or project, they can be very productive.

ADHD can be managed by medication including methylphenidate, more commonly known by the American brand name Ritalin, which improves focus and organisation. It is a controlled drug because it can be abused (it is also known by its street name "speed", an illegal Class B amphetamine) and can lead to addictive behaviours.

Source: Ely, J., "Women in their 20s and 30s drive record surge in ADHD prescriptions amid row over whether condition is being OVER-diagnosed". *Mail Online*, 7July 2023. Available at: https://www.dailymail.co.uk/health/article-12270783/ADHD-prescriptions-jump-33-year-women-20s-30s-driving-record-uptick.html

Autism

Autism affects social communication, interaction and behaviour. For someone to be diagnosed as autistic (under the DSM-5 medical diagnostic guidelines), they need to show that they face ongoing challenges with "social communication and interaction across multiple contexts" and "restricted, repetitive patterns of behaviour, interests or activities". Autistic people may have difficulties with understanding and expressing verbal and non-verbal communication, such as tone, gesture, facial expression and body language.

- Strengths in the workplace may include:
  - Problem-solving
  - Analytical thinking
  - Pattern identification
  - Honesty
  - Reliability

- Challenges in the workplace may include:
  - Dealing with change and transitions
  - Being slower to make sense of what is being said
  - Difficulty in perceiving others' perspectives, thoughts and feelings
  - Difficulty in recognising, understanding, managing and expressing their own emotions
  - Taking things literally when they were meant figuratively
  - Not understanding social norms or cues

Autism presents differently in different people, so a generalisation about how an autistic person might experience the world is unlikely to hold true for any given individual. A supportive approach involves practical techniques that can help reduce autism's unwanted impacts; unlike ADHD, there are no medications that can offer direct help. One such technique is to provide individuals with headphones so that they can block out some of the additional sensory stimulation that is common in an open-plan office.

Numerous articles over the years have claimed that autism is linked to, for example, vaccines, bad parenting or trauma; all of these myths have been debunked.[27] Furthermore, autism is not something that you can "grow out of", as previous myths have contended. While it is now understood that there is a genetic component to autism,[28] research into its causes remains ongoing.

A useful tool in gaining insight into autistic individuals is the so-called "theory of mind" (ToM).[29] ToM refers to the capacity to put yourself in someone else's shoes: to understand that others have different beliefs, behaviours, motivations and thought processes. An individual with autism is likely to differ to a neurotypical individual in this regard. For example, they may struggle to comprehend the motivations that lead another person to lie. ToM is subject to debate among researchers and has some challenges – for example, some would argue that there is a lack of understanding both ways in neurotypical-autistic relationships (the "double empathy" problem[30]); other studies have shown that there is a high degree of mutual understanding within autistic peer groups.[31] While ToM has its challenges, it can be of use in understanding differences in communication and social behaviour between autistic and non-autistic adults.

Neurodivergent individuals, particularly those with ADHD or autism, can find themselves in a state of what is termed "hyperfocus": an intense state of focus, or selective attention.[32] This occurs only when a task is engaging (e.g. fun, interesting, important to the individual), or when the task has purpose in itself. Hyperfocus has been shown to drive an increase in performance and productivity, and a decreased awareness of time and surroundings (e.g. working through lunch and toilet breaks without noticing).

Tourette's and tics

Tourette's and tics are neurological disorders characterised by involuntary and repetitive movements or sounds. Tics can be motor or vocal, simple or complex, and vary in frequency, intensity and duration; they include blinking, jumping, twirling, grunting, throat clearing or coughing. Tics can be influenced by stress, emotions or environmental factors. A diagnosis of Tourette's syndrome might be given if you've had several tics for over a year.[33]

- Strengths in the workplace may include:
  - Creativity
  - High energy levels
  - Being empathetic
  - Being resilient
- Challenges in the workplace may include:
  - Social challenges
  - Stress and anxiety
  - Struggling with tasks that involve physical or verbal control, such as presentations or meetings
  - Being misinterpreted as mocking, disruptive or attention-seeking

Most of the neurodivergent conditions described above are lifelong, but *some* neurodiversity can be acquired or grown out of. Tics and

Tourette's often peak in childhood and adolescence, typically between the ages of 8 and 14, and may decrease as the brain matures. For some, they persist into adulthood but may become less severe or frequent.

Neurodivergence does not present the same in any two individuals. Because of this, in the interest of enhancing communication and understanding, it is more helpful to talk about an individual's specific strengths and challenges/difficulties rather than using medical labels. We will be discussing ways to do that throughout this book.

Many neurodivergent conditions can be considered "invisible": that is to say that, in such a case, you would not necessarily infer that someone had the condition simply by spending time with them.

Example: Chessa

Chessa works in an investment bank, in a fast-paced, high-pressure team. She was diagnosed with autism at a young age, something that the company is aware of. However, her role requires her to interact with a broad range of stakeholders, some of whom are not aware of her differences. Her communication style can be described as "logical", lacking many of the conventional "niceties" of British business communication. For example, she will start an email without a customary "I hope you are well?" (or suchlike), instead getting straight to the heart of the matter in direct and unadorned language. Similarly, she sees small talk at the start of meetings as unnecessary, and moves onto the agenda quickly whenever possible. Some colleagues, unaware of her autism, perceive her to be rude.

As the Chessa example helps to illustrate, many of the traits associated with neurodivergence may not be immediately recognisable by others as being neurodiversity-related. This is why neurodivergence is often described as an "invisible" disability; the potential result of this is a reduced level of empathy and understanding when ND behaviours cause friction in the workplace, and, consequently, increased stress for the person with the condition.[34]

Variances in the definition of neurodivergence

Neurodivergence, as we refer to in this book, is a term applied to variations in brain function and development that have been present from early life. We have listed some of the main neurodivergent conditions above, but there are others, such as Developmental Language Disorder (DLD). Some definitions of neurodivergence are broader than ours, however: these include mental health conditions such as anxiety and depression, which also impact how people experience the world and interact with others. For the purposes of this book, we are taking the narrower definition.

Self-reflection: your expectations

Example: Maya

Maya is a high performer at work who was offered an opportunity to build a relationship with her boss: this would happen by means of a video call. While on the call, she switched the camera off. While it wasn't explicitly stated that cameras were be kept on in relationship-building meetings, there was an implicit expectation that she would have understood the significance of doing so. For Maya, though, a video link doesn't provide any useful information as she finds it difficult to read facial expressions; in fact, it can create sensory overstimulation.

As demonstrated in the example of Maya, cultural and organisational norms exist, and can be explicitly or implicitly stated. Explicit expectations include things like dress codes. Implicit ones can include social taboos surrounding what not to ask someone when you don't know them well, attitudes about social events, or how emotions are expressed. Reflecting on your own expectations in areas like these can make you more aware of how you respond when someone doesn't comply with the norms. As in the Maya example, when norms are not explained, neurodivergent individuals may not pick up on them, leading to confusion.

To identify your own expectations, reflect on the situations below; in each case someone has not done what was expected, or perhaps their behaviour at work has been misunderstood. For each situation, consider:

- What is the expectation?
- Is this expectation explicitly stated or implicitly assumed?
- What examples can you think of in which someone has misunderstood an expectation? What happened in those cases?

1. What clothing is expected in work meetings (i.e. business dress code)?

   ....

   ....

   ....

2. What important days or events are marked or celebrated in your workplace?

   ....

   ....

   ....

3. How do team members greet each other online or in person?

   ....

   ....

   ....

4. How do people indicate that they wish to speak (in online meetings or in person)?

   ....

   ....

   ....

5. What expectations are there around personal space and touch?

   ....

   ....

   ....

6. How is authority treated in your workplace?

   ....

   ....

   ....

7. What are the expectations about arriving on time?

   ....

   ....

   ....

8. How much small talk is acceptable? (If someone skipped small talk, or talked a lot about their own lives, how might it be perceived?)

   ....

   ....

   ....

Neurodiversity and intersectionality

Neurodiversity cannot be understood in isolation from the broader context of intersectionality.

> **Box 2.6: Definition of intersectionality**
>
> Intersectionality is the concept whereby different aspects of a person's identity - such as gender, race, class, sexuality, disability and religion - interact and intersect to create unique and complex experiences of oppression and privilege.

Neurodivergent people may face multiple and overlapping forms of discrimination and disadvantage based on their intersecting identities. Research indicates that these multiple disadvantages are correlated to poor life outcomes and mental health. For example, a black autistic woman may face racism, sexism and ableism, and may have less access than a white autistic man to diagnosis, support and opportunities. Figure 2.2 shows the range of categories that could intersect and thereby cause this compounded bias.

Ability	Nationality	Race	Gender	Gender identity
Culture	Language	Mental health	Age	Class
Appearance	Religion	Education	Neurodiversity	Physical health

Figure 2.2 Neurodiversity and intersectionality

Source: Bayeh, R., *Neurodiversity, Intersectionality and Distress: A Quantitative Survey on the Experiences of University Students*. Master's thesis, Concordia University, 2022.

Gender differences and neurodiversity

As mentioned above, historical stereotypes associated with ADHD and autism have resulted in more men than women being diagnosed. We also saw that increased understanding of both conditions means that more women are now being diagnosed. Discrepancies in gender diagnosis may also be on account of:

- Women masking their symptoms more than men, as a result of being socialised differently from an early age.[35] For example, autistic women may mask their preference to spend time alone (i.e.

force themselves to spend time with people even when that takes a lot of energy and effort for them to do so)

- "Diagnostic overshadowing", which means that doctors typically notice mental health conditions rather than the underlying neurodivergence (e.g. women are diagnosed with anxiety rather than more accurately with ADHD)[36]
- Women are more likely than men to present as the inattentive type of ADHD, which is less visible and less disruptive, and therefore less recognised

Further challenges

We have already explored ways in which the workplace can pose challenges to neurodivergent individuals on account of their differences and the failure by others to acknowledge or accommodate these differences. These challenges mean that neurodivergent individuals are more likely than others to experience:[37]

- **Harassment**. A lack of understanding about, or preconceived biases around, neurodivergent conditions can lead to harassment. For example, some autistic people use "stimming" to help self-regulate their emotions. One form of stimming is the repeated tapping of one's fingers: such behaviour might attract mockery by others in the office
- **Bias**. Individuals may experience bias relating to neurodiversity but also compounded bias relating to neurodivergence and another intersectionality (e.g. race). For example, a bias against people with autism might arise because a previous team member who was autistic struggled with managing his emotions, and the assumption prevails that autistic individuals are all the same
- **Discrimination**. Discrimination may be a result of the way a workplace is structured, for example where recruitment is concerned, whereby implicit exclusion is the unintended outcome of an interview-based (rather than skills-based) process
- **Disclosure**. Often, individuals do not want to disclose their condition to their manager or HR contact within the firm. A US survey

showed that only 59% of neurodivergent people had disclosed their condition at work.[38] Disclosure is often driven by a desire to receive practical and emotional support

- **Mental health**. Mental health challenges are experienced more by neurodivergent individuals than by neurotypical individuals. For example, it is estimated that up to 50% of adults with ADHD suffer from anxiety,[39] and children as young as three years of age with dyspraxia have higher rates of anxiety than those without a diagnosis.[40] Autism has been shown to be associated with higher rates of anxiety, eating disorders, mood disorders and schizophrenia[41]

Case study

> The few colleagues I have confided in have completely dismissed ADHD – 'sure we're all a bit ADHD', 'you've been scrolling TikTok too much' – and I felt belittled and ridiculed. The thought of having to tell more people about my neurodivergence makes me anxious, so I am stuck between a rock and a hard place! Keep it to myself and continue to struggle or disclose this and open myself up to ridicule and discrimination.
>
> 44-year-old female ADHD senior academic

> Once I disclosed my ADHD diagnosis to a manager who immediately said that they couldn't tell I had it. I felt totally invalidated and it was a humiliating experience.
>
> Early-thirties female technology architect with ADHD

These cases demonstrate how preconceived, inaccurate perceptions of what neurodivergence – and specifically ADHD – looks like can cause women to choose not to disclose. Fear of harassment because of social attitudes and expectations, as well detrimental treatment because of their differences, can prevent those with multiple challenges from accessing support – provision of which the workplace might ultimately benefit from.

Key points from Chapter 2

1. The concept of neurodiversity helps us acknowledge the natural diversity in human brains; it shows us that different ways of thinking and behaving are valuable and can be celebrated.
2. There are different models that help people think about neurodiversity. The main ones are the Medical Model, which sees divergence from "normal" brain functioning as something that needs to be fixed, and the Social Model, which sees the barriers that neurodivergent people face as being caused by the environment and society in which they operate. There is currently no agreement about a preferred model.
3. Common neurodivergent conditions include dyslexia, dyspraxia, dysgraphia, OCD, ADHD, autism and Tourette's. Each has its unique characteristics, its degree of impact on the individual and its particular set of workplace challenges and strengths. Often they are not present in isolation, and are not all diagnosed.
4. The "invisibility" of neurodivergent individuals' conditions often creates barriers and biases, which can lead to misunderstandings, isolation and mental health challenges.
5. It is crucial to understand the effect of intersectionality, as neurodivergent people may experience compounded discrimination based on other aspects of their identity such as gender or race.

Notes

1 Miranda-Ojeda, R., Wickramasinghe, A., Ntolkeras, G., Castanho, I., and Yassin, W., "The neurodiversity framework in medicine: On the spectrum". *Journal of Developmental Neurobiology*, 85(1) (2025). https://doi.org/10.1002/dneu.22960

2 Cook, J., Hull, L., Crane, L., et al., "Camouflaging in autism: A systematic review". *Clinical Psychology Review*, November 2021: 89. https://doi.org/10.1016/j.cpr.2021.102080. Bradley, L., Shaw, R., Baron-Cohen, S., and Cassidy, S., "Autistic adults' experiences of camouflaging and its perceived impact on mental health". *Autism in Adulthood*, 3(4) (December 2021). https://doi.org/10.1089/aut.2020.0071. Evans, J., Krumrei-Mancuso, E.J., and Rouse, S.V., "What you are hiding could be hurting you: Autistic masking in relation to mental health, interpersonal trauma, authenticity, and self-esteem". *Autism in Adulthood*

6(2) (2022). https://doi.org/10.1089/aut.2022.0115. Prykes-Hobbs, A., et al., "The workplace masking experiences of autistic, non-autistic neurodivergent and neurotypical adults in the UK". *PLOS One*, September 2023. https://doi.org/10.1371/journal.pone.0290001.

3 Samaha, A.M., "What good is the social model of disability?" *The University of Chicago Law Review*, 74(4) (2007), 1251-308.

4 Institute of Leadership, "Workplace neurodiversity: The power of difference". Institute of Leadership, 2025. Available at: https://leadership.global/resourceLibrary/workplace-neurodiversity-the-power-of-difference.html

5 Parris, M., "Our disability benefits system invites abuse". *The Times*, 8 March 2024. Available at: https://www.thetimes.com/article/54358729-a25f-4fa6-8fd5-07426d3c9a6c

6 For example: Boyle, S., "The sudden rise of AuDHD: What is behind the rocketing rates of this life-changing diagnosis?" *The Guardian*, 4 April 2024. Available at: https://www.theguardian.com/lifeandstyle/2024/apr/04/audhd-what-is-behind-rocketing-rates-life-changing-diagnosis. Tsjeng, Z., "What is ADHD, how do you get a diagnosis and can you only treat it with drugs? All your questions, answered by experts". *The Guardian*, 29 November 2024. Available at: https://www.theguardian.com/lifeandstyle/2024/nov/29/what-is-adhd-how-do-you-get-a-diagnosis-and-can-you-only-treat-it-with-drugs-all-your-questions-answered-by-experts. Roper, M.F., "Dyspraxia can be serious: It deserves more recognition". *The Guardian*, 29 October 2012. Available at: https://www.theguardian.com/commentisfree/2012/oct/29/dyspraxia-serious-recognition

7 Connolly, S.E., Constable, H.L., and Mullally Sinéad L., "School distress and the school attendance crisis: A story dominated by neurodivergence and unmet need". *Frontiers in Psychiatry*, 14 (2023). https://doi.org/10.3389/fpsyt.2023.1237052. National Autistic Society, *School Report 2021*. National Autistic Society, 9 November 2021. Available at: https://www.autism.org.uk/what-we-do/news/school-report-2021. Graham, B., et al., *School Exclusions: A Literature Review on the Continued Disproportionate Exclusion of Certain Children*. Department for Education, May 2019. Available at: https://assets.publishing.service.gov.uk/media/5cd15de640f0b63329d700e5/Timpson_review_of_school_exclusion_literature_review.pdf. Ringbom, I., et al., "Psychiatric disorders diagnosed in adolescence and subsequent long-term exclusion from education, employment or training: Longitudinal national birth cohort study". *British Journal of Psychiatry: The Journal of Mental Science*, 220(3) (March 2022), 148–53. https://doi.org/10.1192/bjp.2021.146

8 Reindal, S.M., "A social relational model of disability: A theoretical framework for special needs education?" *European Journal of Special Needs Education*, 23(2) (2008), 135–46. https://doi.org/10.1080/08856250801947812. Doyle, N., "Neurodiversity at work: A biopsychosocial model and the impact on working adults". *British Medical Bulletin* 135 (1) (September 2020), 108–25. https://doi.org/10.1093/bmb/ldaa021. Whelpley, C.E., Holladay-Sandidge, H.D., Woznyj, H.M. and Banks, G.C., "The biopsychosocial model and neurodiversity: A person-centered approach. *Industrial and Organizational Psychology*, 16(1) (9 March 2023), 25–30. https://doi.org/10.1017/iop.2022.95.

9 Madigan, N., "More people are being diagnosed with neurodiversity than ever. Here's why it's a good thing". MSN, 6 June 2024. Accessed February 2025 at: https://www.msn.com/en-au/health/other/more-people-are-being-diagnosed-with-neurodiversity-than-ever-heres-why-its-a-good-thing/ar-BB1nHXJG.

10 Connolly, S.E., Constable, H.L., and Mullally Sinéad L., "School distress and the school attendance crisis: A story dominated by neurodivergence and unmet need". *Frontiers in Psychiatry*, 14 (2023). https://doi.org/10.3389/fpsyt.2023.1237052. National Autistic Society, *School Report 2021*. National Autistic Society, 9 November 2021. Available at: https://www.autism.org.uk/what-we-do/news/school-report-2021. Graham, B., et al., *School Exclusions: A Literature Review on the Continued Disproportionate Exclusion of Certain Children*. Department for Education, May 2019. Available at: https://assets.publishing.service.gov.uk/media/5cd15de640f0b63329d700e5/Timpson_review_of_school_exclusion_literature_review.pdf. Ringbom, I., et al., "Psychiatric disorders diagnosed in adolescence and subsequent long-term exclusion from education, employment or training: Longitudinal national birth cohort study". *British Journal of Psychiatry: The Journal of Mental Science*, 220(3) (March 2022), 148–53. https://doi.org/10.1192/bjp.2021.146

11 Fairclough, M., "Can't, not won't: Rethinking school avoidance". *British Psychological Society*, 27 November 2024. Available at: https://www.bps.org.uk/psychologist/cant-not-wont-rethinking-school-avoidance.

12 Black, R.H., *A Narrative Enquiry into the Experiences of Women with a Late Diagnosis of Attention Deficit Hyperactivity Disorder (ADHD)*. Unpublished doctoral thesis, University of Chester, 2023. Madigan, N., "More people are being diagnosed with neurodiversity than ever. Here's why it's a good thing". MSN, 6 June 2024. Accessed February 2025 at: https://www.msn.com/en-au/health/other/more-people-are-being-diagnosed-with-neurodiversity-than-ever-heres-why-its-a-good-thing/ar-BB1nHXJG. Russell, G., Stapley, S., Newlove-Delgado, T., Salmon, A., White, R., Warren, F., Pearson, A. and Ford, T., "Time trends in autism diagnosis over 20 years: A UK population-based cohort study". *Journal of Child Psychology and Psychiatry*, 63(6) (2021), 674–82, https://doi.org/10.1111/jcpp.13505. National Institute for Health and Care Research, "Significant rise in ADHD diagnoses in the UK". NIHR, 18 July 2023. Available at: https://www.nihr.ac.uk/news/significant-rise-adhd-diagnoses-uk

13 Martin, J., "Why are females less likely to be diagnosed with ADHD in childhood than males?" *The Lancet Psychiatry* 11(4) (April 2024), 303–10. https://doi.org/10.1016/S2215-0366(24)00010-5. Cook, J., Hull, L. and Mandy, W., "Improving diagnostic procedures in autism for girls and women: A narrative review". *Neuropsychiatric Disease and Treatment*, 20 (2024), 505–14. https://doi.org/10.2147/NDT.S372723.

14 Huang-Horowitz, N., Peterson, H., and Furey, L., "Exploring the media discourse of neurodiversity and neurodivergent individuals: A step toward discovering possibilities for all people". *Western Journal of Communication*, 89(4) (2025), 654–74. https://doi.org/10.1080/10570314.2024.2395314. Shayla, L., "The neurodiversity movement has a moment this year", *Discover*, 7 December 2023. Available at: https://www.discovermagazine.com/mind/

the-neurodiversity-movement-has-a-moment-this-year. BBC, "Apology issued from NHS Borders over ADHD diagnosis delay". BBC, 19 July 2023. Available at: https://www.bbc.co.uk/news/uk-scotland-south-scotland-66244454. BBC, "NHS Tayside halts new ADHD and autism referrals to CAMHS". BBC, 20 March 2025. Available at: https://www.bbc.co.uk/news/articles/cy5nv744ek7o

15 Examples from 2025: Del Fabbro, L., "Comedian Sue Perkins says she does not 'weaponise' her ADHD to make excuses". *Evening Standard*, 8 April 2025. Available at: https://www.msn.com/en-us/health/other/comedian-sue-perkins-says-she-does-not-weaponise-her-adhd-to-make-excuses/ar-AA1CtINZ?ocid=BingNewsSerp. McLaughlin, C., "Aimee Lou Wood says she was diagnosed with ADHD and autistic traits". *Evening Standard*, 6 April 2025. Available at: https://www.msn.com/en-gb/health/other/aimee-lou-wood-says-she-was-diagnosed-with-adhd-and-autistic-traits/ar-AA1CnV6g?ocid=BingNewsSerp. *Additude* Editors, "Famous people with ADHD". *Additude*, 19 March 2025. Available at: https://www.additudemag.com/slideshows/famous-people-with-adhd

16 Voltaire, S., Steinberg, H., Garfield, T., Chvasta, K., Ardeleanu, K., Brown, M. and Shea, L., "Inextricably tied: Nonbinary autistic individuals' views on how their gender identity and autism are connected". *Autism*, 28(12) (2024), 3156–66. https://doi.org/10.1177/13623613241257600. Goetz, T.G. and Adams, N., "The transgender and gender diverse and attention deficit hyperactivity disorder nexus: A systematic review". *Journal of Gay and Lesbian Mental Health* 28(1), (2024), 2–19. https://doi.org/10.1080/19359705.2022.2109119.

17 Chellappa, S.L., "Addressing multiple neurodivergent identities in clinical and research settings". *The Lancet Child & Adolescent Health*, 9(1) (2025), 5–6. https://doi.org/10.1016/S2352-4642(24)00261-X. Leitner, Y., "The co-occurrence of autism and attention deficit hyperactivity disorder in children: What do we know?" *Frontiers in Human Neuroscience*, 8 (April 2014), 268. https://doi.org/10.3389/fnhum.2014.00268

18 British Dyslexia Association, "About dyscalculia". British Dyslexia Association, n.d. Available at: https://www.bdadyslexia.org.uk/dyscalculia/how-can-i-identify-dyscalculia

19 British Dyslexia Association, "Dyslexia". British Dyslexia Association, n.d. Available at: https://www.bdadyslexia.org.uk/dyslexia

20 Chung, P.J., Patel, D.R. and Nizami, I., "Disorder of written expression and dysgraphia: Definition, diagnosis and management". *Translational Paediatrics*, 9(1) (2020), S46–S64. https://doi.org/10.21037/tp.2019.11.01

21 Meachon, E., "An investigation of dyspraxia: What we know and why the research is so far behind". *ResearchGate*, October 2017. Available at: https://www.researchgate.net/publication/321275661_An_Investigation_of_Dyspraxia_What_We_Know_and_Why_the_Research_Is_So_Far_Behind

22 O'Nions, E., El Baou, C., John, A., *et al.*, "Life expectancy and years of life lost for adults with diagnosed ADHD in the UK: Matched cohort study". *British Journal of Psychiatry*, 2025, 1–8. https://doi.org/10.1192/bjp.2024.199

23 British Medical Association, "Autism spectrum disorder". BMA, 2024. Available at: https://www.bma.org.uk/what-we-do/population-health/improving-the-health-of-specific-groups/autism-spectrum-disorder

24 OCD UK, "Occurrences of OCD". OCD UK, n.d. Available at: https://www.ocduk.org/ocd/how-common-is-ocd

25 Tourette's Action, "#TourettesWeMatter". Tourette's Action, 2025. Available at: https://www.tourettes-action.org.uk/150-touretteswematter.html

26 American Psychiatric Association, *Diagnostic and Statistical Manual of Mental Disorders*, 5th edn, Text Revision (DSM-5-TR). American Psychiatric Association, 2013.

27 Public Health on Call, "Vaccines don't cause autism: Why do some people think they do?" Johns Hopkins University Bloomberg School of Public Health, 19 March 2025. Available at: https://publichealth.jhu.edu/2025/the-evidence-on-vaccines-and-autism. Cox, D., "The genetic mystery of why some people develop autism". BBC Future, 16 April 2025. Available at: https://www.bbc.co.uk/future/article/20250415-the-genetic-mystery-of-why-some-people-develop-autism

28 Cox, D., "The genetic mystery of why some people develop autism". BBC Future, 16 April 2025. Available at: https://www.bbc.co.uk/future/article/20250415-the-genetic-mystery-of-why-some-people-develop-autism. Thapar, A., and Rutter, M. "Genetic advances in autism". *Journal of Autism and Developmental Disorders*, 51(12) (December 2021), 4321–32. https://doi.org/10.1007/s10803-020-04685-z.

29 Baron-Cohen, S., *Mindblindness: An Essay on Autism and Theory of Mind*. Boston, MA: MIT Press/Bradford Books, 1995.

30 Milton, D. E., "On the ontological status of autism: The 'double empathy problem'". *Disability & Society, 27* (6) (2012), 883-887. https://doi.org/10.1080/09687599.2012.710008

Milton, D., Gurbuz, E., & López, B. "The 'double empathy problem': Ten years on". *Autism: The International Journal of Research and Practice*, 26(8) (2022), 1901–03. https://doi.org/10.1177/13623613221129123. Holt, A., Bounekhla, K., Welch, C. and Polatajko, H., "Unheard minds, again and again": autistic insider perspectives and theory of mind. *Disability and Rehabilitation, 44* (20) (2022), 5887–97. https://doi.org/10.1080/09638288.2021.1949052

31 Crompton, C. J., Hallett, S., Ropar, D., Flynn, E. and Fletcher-Watson, S., "I never realised everybody felt as happy as I do when I am around autistic people": A thematic analysis of autistic adults' relationships with autistic and neurotypical friends and family. *Autism: The International Journal of Research and Practice, 24* (6) (2020), 1438–48. https://doi.org/10.1177/1362361320908976

32 Ashinoff, B.K., and Abu-Akel, A., "Hyperfocus: The forgotten frontier of attention". *Psychological Research*, 85(1) (February 2021), 1–19. https://doi.org/10.1007/s00426-019-01245-8

33 NHS, "Tourette's syndrome". NHS, n.d. Available at: https://www.nhs.uk/conditions/tourettes-syndrome

34 Abney, A., Denison, V., Tanguay, C. and Genz, M., "Understanding the unseen: Invisible disabilities in the workplace". *The American Archivist*, 85(1) (2022): 88-103. Available at: https://scholarcommons.sc.edu/cgi/viewcontent.cgi?article=1057&context=lib_facpub. Hicksted, J.L., "Stigma associated with invisible disabilities and its effect on intended disclosure in the workplace". Walden University ProQuest Dissertations and Theses, 2023.

35 Black, R.H., *A Narrative Enquiry into the Experiences of Women with a Late Diagnosis of Attention Deficit Hyperactivity Disorder (ADHD)*. Unpublished doctoral thesis, University of Chester, 2023.

36 Agnew-Blais, J.C., "Hidden in plain sight: Delayed ADHD diagnosis among girls and women – a commentary on Skoglund et al. (2023)". *Journal of Child Psychology and Psychiatry*, 65(10) (October 2024), 1398–400. https://doi.org/10.1111/jcpp.14023. Gupta, N., and Gupta, M., "Diagnostic overshadowing in high-functioning autism: Mirtazapine, Buspirone, and modified cognitive behavioral therapy (CBT) as treatment options". Cureus, 15(5) (24 May 2023), e39446. https://doi.org/10.7759/cureus.39446

37 CIPD, *Neuroinclusion at Work Report 2024*. CIPD, 2024. Available from: https://www.cipd.org/globalassets/media/knowledge/knowledge-hub/reports/2024-pdfs/2024-neuroinclusion-at-work-report-8545.pdf

38 McGlauflin, P., and Burleigh, E., "Neurodivergent employees aren't disclosing their conditions at work because they worry it will hurt their career". *Fortune*, 15 May 2024. Available at: https://fortune.com/2024/05/15/neurodivergent-employees-arent-disclosing-their-conditions-at-work-because-they-worry-it-will-hurt-their-career

39 Katzman, M.A., Bilkey, T.S., Chokka, P.R., et al., "Adult ADHD and comorbid disorders: Clinical implications of a dimensional approach". *BMC Psychiatry*, 17(302) (August 2017). https://doi.org/10.1186/s12888-017-1463-3

40 Piek, J.P., Bradbury, G.S., Elsley, S.C. and Tate, L., "Motor coordination and social-emotional behaviour in preschool-aged children". *International Journal of Disability, Development and Education*, 55(2) (2008), 143–51. https://doi.org/10.1080/10349120802033592

41 Zheng, Z., Zheng, P., and Zou, X., "Association between schizophrenia and autism spectrum disorder: A systematic review and meta-analysis". *Autism Research*, 11(8) (2018), 1110–19. https://doi.org/ 10.1002/aur.1977. Gotham, K., Bishop, S.L., Hus, V., Huerta, M., Lund, S., Buja, A., Krieger, A., and Lord, C., "Exploring the relationship between anxiety and insistence on sameness in autism spectrum disorders". *Autism Research*, 6(1) (2012), 33–41. Oakley, B., Loth, E., and Murphy, D.G., "Autism and mood disorders". *International Review of Psychiatry*, *33* (3) (2021), 280–99. https://doi.org/10.1080/09540261.2021.1872506

3 Shifting biases

Identifying your current bias

Chapter 2 had an exercise to help you to identify your current neurodiversity-related understanding, assumptions and biases (such as the belief that all autistic people have a high IQ). However, we all have *unconscious* assumptions and thought patterns that influence our thoughts and behaviour without our being aware of them. In this chapter, we will discuss different kinds of unconscious thought patterns, some of which may be classed as biases - unconscious tendencies that skew judgement and often result in unfair treatment, inaccuracies or distortions. We will also discuss how biases can become institutionalised, thereby creating environments that are skewed: specifically, how they can be skewed against neurodivergent individuals.

Where does bias come from?

Bias is inherent in the way we think, and it derives from the way our brains necessarily create mental shortcuts in order to process information efficiently.

According to psychologist Daniel Kahneman,[1] our brain operates a two-tier system. When we need to act fast, our brain uses "System 1": quick thinking that uses limited information and bases decisions on intuition, instinct and assumptions. This system enables us to be ready for action, using previous experiences to inform how we might respond. For example, System 1 thinking is what happens when you take a shower at home. Because you have done it many times before,

DOI: 10.4324/9781003645535-3

Table 3.1 Fast versus slow thinking

System 1 Thinking ("Fast")	*System 2 Thinking ("Slow")*
Unconscious, automatic	Deliberate, concerted
Generally based on emotion and embedded memories	Based mainly on logical and analytical thinking
Efficient and requiring low energy consumption	Requires more cognitive energy
With the aim of keeping you safe	Often driven by an individual's core values, such as respect
Can be biased	Can help challenge bias through increased awareness of, and reflection on, behaviours that were unconscious

you know where everything is and you don't have to actively think about what tasks you are doing. You're on autopilot. Unless someone has changed the water temperature since you were last in the shower, or someone has moved your shower gel, you don't have to make any changes since the last time you did it.

System 1 is also responsible for making quick assessments of situations, which can, for example, get you out of danger quickly. For instance, if you're walking along a poorly lit alleyway, you might assume the person behind you is a threat and react by walking faster. This may be based on previous knowledge or experiences that have shown that people in alleyways can be threatening or cause trouble. You need a quick answer to a common problem and can't spend time thinking about options or performing a more thorough evaluation.

A lot of these shortcuts are created when we are very young, based on experiences with family, friends and the world around us. We are often not aware of them, as they are embedded and absorbed by observing how our carers behave. From a neuroscience perspective, we are creating neural pathways to form these automatic reactions. These neural pathways have been laid down and then reiterated through years of use; each use of the same thought pattern makes it a more well-worn and an easier path to take. One workplace example of this is a manager who, without thinking about it, always assigns high-profile tasks to people that are the same (gender, ethnicity), despite having a diverse and capable team.

System 2 thinking, by contrast, is slower, deliberate and focuses on details. It is the system that you use when you want to learn something new and therefore create new neural pathways. For example, using new computer software for the first time takes concentration and thinking capacity (i.e. System 2). To give another example: when you start a new role at work, you need to take in new information and remember it: the System 2 thinking required for this is what makes those first few months in a new role more tiring.

The energy-intensiveness of System 2 means we cannot use it all the time; instead we use the shortcuts in System 1. When these shortcuts have a negative impact on others, we might call them "biases".

Box 3.1: Conscious versus unconscious bias

A bias is a tendency, inclination or prejudice that results in unfair judgement, unequal treatment or a distorted perception.

Conscious bias, also known as explicit bias, refers to the attitudes or beliefs that individuals hold consciously and express openly. These biases are deliberate and intentional, and people who possess them are fully aware of their views.

Unconscious bias, also known as implicit bias, refers to those attitudes or stereotypes that influence our understanding, actions and decisions unknowingly.

Using the reflective exercises in this chapter can help you become more aware of your own predictive, quick, assumptive thinking about a situation, person or behaviour. Becoming aware of how your brain operates is the first step towards becoming a more inclusive manager and leader.

Box 3.2: But I'm not biased!

Many people don't accept that they have significant unconscious biases. To challenge yourself on this, try a test called the Implicit

Association Test.[2] Designed by researchers at Harvard University, it identifies how you associate certain words. For example, you mayfeel a stronger association between the words "man" and "leader", than you do between "woman" and "leader". There currently isn't a test specifically for neurodiversity-related bias, but undertaking this test can help raise your awareness about unconscious bias more generally.

Another useful exercise is to reflect on what messages you received as you were growing up. How did your family and friends talk about other groups of people? How has that shaped what you think now?

Cognitive biases

Having accepted that these cognitive biases, as they are called, are a natural part of having a human brain, let's focus on the common cognitive biases that are most likely to impact neurodivergent individuals.

Confirmation bias

This shortcut saves our brain's energy by filtering out information that doesn't fit with our pre-existing world-view but instead actively seeks data that fits with our current thinking. For example, a manager may have a perception that autistic individuals are geeky males who are very intelligent. When they meet someone who doesn't conform to that - say, a middle-aged woman of average intelligence - they might find it difficult to believe that she is autistic, because it doesn't fit their view.

You can challenge your confirmation bias by:

- Actively looking for data or facts that are the exception, or differ from what you already know. Push yourself to consider what you can take from this new information and incorporate into your world-view
- Inviting colleagues, in discussions, to adopt different perspectives; allow them to speak first before you (as their leader) disclose your own priorities and motivations. You may even assign "devil's

advocates", whose express role is to challenge your perspective and assumptions

Conformity bias

Also known as peer pressure, this is the tendency to align your opinions and behaviours with those of a group in order to fit in and be accepted. This bias can result in suboptimal outcomes, because fewer options and opinions are shared and a diversity of perspectives is lacking. For instance, an employee in a team meeting might have information that would expose the unworkability of a plan of action. But they agree with the plan nonetheless because their manager proposed it and they don't want to appear uncooperative or difficult.

Liking bias

If you've ever bought something because you liked the person selling it, even though you didn't need what they were selling, then you've been subject to liking bias. The more you like someone, the more inclined you are to buy from them, be influenced by them or want to help them. Research has shown that we perceive people to be "likeable" if:

1. We find them attractive to look at
2. They are similar to us in terms of interests, origin or personality
3. We believe they like us

While it is not true of all neurodivergent individuals, research indicates that some of the social traits of neurodivergent individuals can make them less likeable compared to neurotypicals,[3] more stigmatised[4] (even when they have not disclosed their condition) and that disabilities can negatively impact relationships between leaders and their teams.[5] For example, autistic people may appear to be abrupt in conversation as a result of their logical responses to questions or their blunt honesty. Autistic people can also have "special interests" - intense fascination in specific areas - which can be difficult for others to relate to.

Awareness of this is all the more critical because people are more likely to give higher performance ratings, or provide sponsorship or mentorship, to those who share similarities to them.[6] Furthermore, the dissimilar individuals are hence more prone to exclusion, isolation and critical feedback. This can be amplified further if their neurological differences are combined with other differences like gender or ethnicity (see Chapter 1 and intersectionalities).

Box 3.3: How bias shows up in the workplace

> I have been criticised by a manager for my facial expressions in meetings, being told I am condescending. This wasn't how I was feeling at all. Autism means that my facial expressions often bear no resemblance to my emotions, but the criticism made me so self-conscious that I tried to hold a slight smile in all meetings after that, which coincided with lockdown and all meetings moving online. I caused myself a twitch by my right eye from holding an unnatural facial pose.
>
> 49-year-old female senior manager
> in professional services

This example shows how social expectations create a requirement for neurodivergent individuals to act a certain way – a way that is not natural for them. Changing their behaviour to adapt to people's expectations allows them to be more "likeable", but can often have unintended consequences.

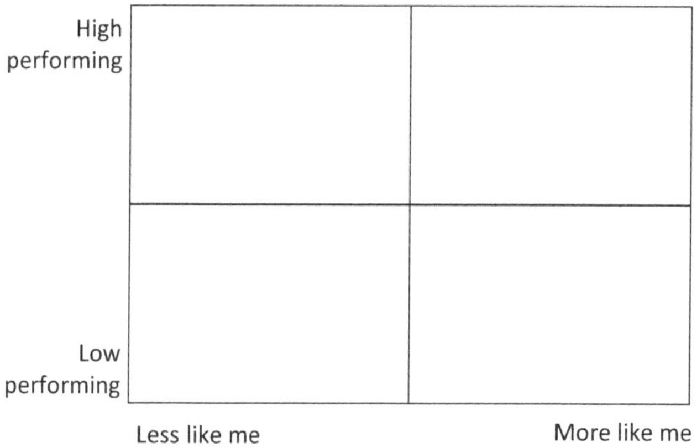

Figure 3.1 Matrix: perceived performance versus similarity

You can challenge your own liking bias by mapping, on the matrix in Figure 3.1, the ten people you most often interact with at work. Define performance in terms of how competent you believe they are in their role; define similarity as how alike they are to you in terms of attractiveness, gender, age, neurotype and background. Do you notice any patterns emerging?

What patterns are emerging, if any?

....

....

....

....

How might your body language differ when you're working with people who you like less, or are more different to?

....

....

....

....

Status quo bias

This is a preference for the current state of affairs or situation, manifesting as resistance to change. This bias can hinder innovation and adaptability. For example, a company might continue using a certain type of technology because individuals are more familiar with it, despite available advancements that could significantly improve efficiency. There is a perception that "what is observable is good", which prevents people from considering what could be better.

For example, although existing research and observations suggest that autistic individuals succeed in technical fields, and that dyslexia is associated with entrepreneurship, such views also become entrenched as societal biases. And so the stereotype of an "autistic tech geek", for example, might prevent a recruiter from considering an autistic individual for a non-technical career.

Proximity bias

This is the tendency to prefer people or things that are physically closer to you. This is particularly worth acknowledging in hybrid workplaces, where people who are there in the office may be viewed more favourably than those who work from home: a manager might assign work to people who are in the office or favour the opinions of those who are in physical meetings rather than virtual participants. The shift towards a greater acceptance of working from home has enabled many neurodivergent individuals to work more effectively; increased prevalence and impact of proximity bias may therefore be an unwanted outcome of this.

Institutionalised bias

What we have uncovered so far are internal biases – unconscious and conscious – that dictate how we operate as people. However, bias can also work at an institutional level, for example when a workplace environment is structured for the benefit of the majority rather than for all.

As an example, some workplace cultures regularly use figurative or metaphorical speech, which many autistic people find confusing. Phrases such as "let's circle back on that" or talking about "smashing it out of the park" can be confusing for people who process words literally.

Another example is workplace design. The last few decades have seen a move in most cultures towards open-plan offices: they are seen as a way of increasing collaboration and creating a more egalitarian workplace. However, this assumes that everyone is comfortable with the resulting noise, lighting and people moving around. While this works for the majority, it can be challenging for some neurodivergent individuals: the distractions and additional sensory input can make it difficult for them to focus, reducing the energy they have available for the tasks in hand.

Institutionalised bias can be embedded in technology. Although tech can reduce bias, it can also enhance it, too. This can show up in many ways; here are just some examples:

- **Design**. Many software applications are designed without neurodivergence in mind. They might have cluttered, overwhelming interfaces, for example, or unclear instructions, making their platforms difficult to use

- **Data sets**. AI (artificial intelligence) systems can inadvertently perpetuate biases. For example, they can filter out the résumés of neurodivergent candidates from a hiring process because they draw on historic (predominantly neurotypical) recruitment patterns as a basis
- **Rollout**. Some companies fail to consider how new platforms or tools might work with accessibility software, such as voice-to-text and text-to-voice

The resources section of this book has suggestions for further reading in this space. Whether you design IT systems or not, the key message is to keep neurodiversity in mind at every stage of any workplace change: design, development and rollout.

What makes this even more important for neurodivergent individuals

Both neurotypical and neurodivergent people can be targets of bias, but studies have shown that the latter can have a different, more heightened, emotional response to their experience of bias.[7]

There are different possible explanations for this. Neurodivergent individuals have been shown to be more likely to have experienced more trauma than neurotypicals,[8] meaning the autonomic nervous system is more easily triggered into a fight-or-flight response to a threat - in this case a biased individual. More specifically, ADHD individuals may be more sensitive to a perception of being rejected (as a result of the bias): a condition called rejection sensitivity dysphoria (see Box 3.4).

Box 3.4: Rejection sensitivity dysphoria

RSD-often associated with ADHD-is an intense emotional reaction to the perception of being rejected by others. It can be triggered by other people's behaviours, perceived criticism or judgement, even when their intentions were good. You may notice it happening: a neurodivergent person might display emotions such as distress, shame, embarrassment, upset and anger. It can lead to rumination and an inability to move on from an upsetting situation.

Note that this is not a diagnosable medical condition, but a common experience of those with ADHD.

Applying this to you

Think of a time when you consider yourself to have been treated unfairly. Maybe you weren't promoted when you thought that you were ready? Perhaps you received feedback that wasn't a fair reflection of your work?

What led to that situation? Who were you working with and what was the background to the situation?

....

....

....

....

Having reflected on what you have learned in this chapter, what biases might have been at play?

....

....

....

....

How did that impact you? At the time? And now?

....

....

How has your perception of the situation changed with time?

Think of a time when you consider yourself to have been excluded. Maybe you weren't invited to an event when you were expecting to be? Perhaps your opinions weren't welcomed or invited in a meeting?

What happened? What was the context?

What feelings/emotions did you experience at the time?

What did you do? For example, did you withdraw, confront or seek comfort?

How do you think about the situation now?

....

....

....

....

What are you learning about yourself?

....

....

....

....

How to shift your bias around neurodiversity

The first step towards any kind of change is to raise your own awareness of what is currently happening. To help you in doing that, answer the questions in Table 3.2, which ask you to reflect on where you have seen bias in your workplace – for you or against you, or towards others. Having done that, look at the questions that follow:

Table 3.2 Identifying biases in our lives

Area	*Description*	*How does this show up for me in my workplace?*
Confirmation bias	Filtering out information that doesn't fit with our pre-existing world-view	
Conformity bias	Aligning one's opinions and behaviours with those of a group in order to fit in or be accepted	
Liking bias	The more you like someone, the more inclined you are to side with or help that person	
Status quo bias	A preference for the current situation or state of affairs, manifesting as resistance to change	
Institutionalised bias	The standard ways of doing things in institutions that are biased against a minority.	

How might the biases listed in Table 3.2 impact those who think differently to you?

....

....

....

....

Having identified biases showing up in your own workplace, what might you do differently now?

....

....

....

....

People can often find it difficult to change their perspectives, as it means relinquishing a part of themselves: a long-held belief, value or even a facet of their identity. For example, those who believe that "their" group (e.g. neurotypical) are better than "the others" (e.g. neurodivergent) will lose that bias only when they let go of a long-held underlying belief around who "the others" are and about their behaviours. Identity can work at a group level: resistance may be encountered when a team built around similar people is asked to include those that are different to them. In such cases, there is a need to grieve for the part of yourself or your group that might be lost before the future version can be fully adopted. This can take time, and possibly repeated attempts, but it is more easily accomplished when you can identify what it is that you are losing and what that means to you.

What might you believe you are losing or giving up as a result of reducing your biases?

....

....

....

....

Are these things important to you? If so, what makes them important?

....

....

....

....

What emotions or feelings do you experience when thinking about this?

....

....

....

....

We will explore more about change and emotions in Chapter 8.

What to do when you notice bias in yourself

Here are some simple steps to help tackle biases you notice in yourself:

1. Clearly define the bias that you are holding. When does it come up (e.g. in what context and with which people)? Who does it impact?
2. Challenge the validity of the idea. To what extent is it based on fact? To what extent is it not based on anything? Is there a past experience on which it might be based?
3. Challenge yourself to change your behaviour. The aim is to change your perspective about the situation. For example, you might:

    a. Perspective-take. Spend time thinking about the experience of the other person. What does their day-to-day look like? What might their childhood have been like? What might be the pressures and joys of their day-to-day life?
    b. Get curious. Ask questions of whoever you think you are biased against. This can help challenge your assumptions about who they are and what they are like
    c. Identify where you are using stereotypes. Is it a valid stereotype? Is there an alternative viewpoint that might be more helpful?

d. Use mindfulness techniques to increase your awareness of your thoughts in the moment, so you can identify your biases showing up in real time and challenge them there and then (see the resources in Chapter 12)

Key points from Chapter 3

1. Our brains are naturally wired to create shortcuts. These are both helpful and unhelpful to us in our day-to-day lives.
2. Sometimes these shortcuts manifest as biases, resulting in unfair judgement, unequal treatment or a distorted perception of situations or people.
3. These biases can be cognitive (e.g. confirmation bias) or embedded into our environment over time (e.g. technological bias).
4. Challenging these biases can lead to a fairer, more inclusive workplace.
5. The first step in challenging biases around neurodiversity is to identify them in yourself.

Notes

1 Kahneman, D., *Thinking Fast and Slow*. New York: Farrar, Straus & Giroux, 2011.
2 Harvard University, "The Implicit Association Test". Harvard University, n.d. Available at: https://implicit.harvard.edu/implicit/
3 Alkhaldi, R.S., Sheppard, E., Burdett, E., and Mitchell, P., "Do neurotypical people like or dislike autistic people?". *Journal of Autism in Adulthood*, 3(3) (September 2021), 275-79. https://doi.org/10.1089/aut.2020.0059. Hoza, B., "Peer functioning in children with ADHD". *Journal of Pediatric Psychology*, 32(6) (July 2007), 655-63. https://doi.org/10.1093/jpepsy/jsm024
4 Lebowitz, M.S., "Stigmatization of ADHD: A developmental review". *Journal of Attention Disorders*, 20(3) (2013), 199-205. https://doi.org/10.1177/1087054712475211. Schrevel, S.J.C., Dedding, C., Van Aken, J.A., and Broerse, J.E.W., "'Do I need to become someone else?' A qualitative exploratory study into the experiences and needs of adults with ADHD". *Journal of Health Expectations*, 19(1) (February 2016), 39-48. https://doi.org/10.1111/hex.12328
5 Colella, A., and Varma, A., "The impact of subordinate disability on leader-member exchange relationships". *Academy of Management Journal*, 44(2) (2001), 304-15. http://www.jstor.org/stable/3069457
6 Fandt, P.M., Labig, C.E., and Urich, A.L., "Evidence and the liking bias: Effects on managers' disciplinary actions". *Employee Responsibility and Rights Journal*, 3 (December 1990), 253-65. https://doi.org/10.1007/BF01384932. Sutton, A.W., Baldwin, S.P., Wood, L., and Hoffman, B.J., "A meta-analysis of the relationship

between rater liking and performance ratings". *Human Performance*, 26(5) (November 2013): 409–29. https://doi.org/10.1080/08959285.2013.836523

7 Hoover, D.W., "The effects of psychological trauma on children with autism spectrum disorders: A research review". *Review Journal of Autism and Developmental Disorders*, 2 (2015): 287–99. https://doi.org/10.1007/s40489-015-0052-y. Lobregt-van Burren, E., Hoekert, M. and Sizoo, B., "Autism, adverse events and trauma". In Grabrucker, A.M., *Autism Spectrum Disorders.* Netherlands: Exon Publications, 2021. Szymanski, K., Sapanski, L. and Conway, F., "Trauma and ADHD: Association or diagnostic confusion? A clinical perspective". *Journal of Infant, Child, and Adolescent Psychotherapy*, 10(1) (2011): 51–59. https://doi.org/10.1080/15289168.2011.575704. Alexander-Passe, N., "Investigating Post Traumatic Stress Disorder (PTSD) triggered by the experience of dyslexia in mainstream school education?" *Journal of Psychology and Psychotherapy*, 5(6) (January 2015). https://doi.org/10.4172/2161-0487.1000215. Wadsworth, L.P., Van Kirk, N., August, M., et al., "Understanding the overlap between OCD and trauma: Development of the OCD trauma timeline interview (OTTI) for clinical settings". *Current Psychology*, 42 (2023), 6937–47. https://doi.org/10.1007/s12144-021-02118-3. Van der Kolk, B.A., "Clinical implications of neuroscience research in PTSD". In *Annals of the New York Academy of Sciences* 1071(1): *Psychobiology of Posttraumatic Stress Disorder: A Decade of Progress* (July 2006). Van der Kolk, B.A., "The neurobiology of childhood trauma and abuse". *Child and Adolescent Psychiatric Clinics*, 12(2) (2003), 293–317. https://doi.org/10.1016/s1056-4993(03)00003-8. Van der Kolk, B.A., "Posttraumatic therapy in the age of neuroscience". *Psychoanalytic Dialogues*, 12(3) (2002), 381–92. https://doi.org/10.1080/10481881209348674.

8 Hoover, D.W., "The effects of psychological trauma on children with autism spectrum disorders: A research review". *Review Journal of Autism and Developmental Disorders*, 2 (2015): 287–99. https://doi.org/10.1007/s40489-015-0052-y. Lobregt-van Burren, E., Hoekert, M. and Sizoo, B., "Autism, adverse events and trauma". In Grabrucker, A.M., *Autism Spectrum Disorders.* Netherlands: Exon Publications, 2021. Szymanski, K., Sapanski, L. and Conway, F., "Trauma and ADHD: Association or diagnostic confusion? A clinical perspective". *Journal of Infant, Child, and Adolescent Psychotherapy*, 10(1) (2011): 51–59. https://doi.org/10.1080/15289168.2011.575704. Alexander-Passe, N., "Investigating Post Traumatic Stress Disorder (PTSD) triggered by the experience of dyslexia in mainstream school education?" *Journal of Psychology and Psychotherapy*, 5(6) (January 2015). https://doi.org/10.4172/2161-0487.1000215. Wadsworth, L.P., Van Kirk, N., August, M., et al., "Understanding the overlap between OCD and trauma: Development of the OCD trauma timeline interview (OTTI) for clinical settings". *Current Psychology*, 42 (2023), 6937–47. https://doi.org/10.1007/s12144-021-02118-3. Van der Kolk, B.A., "Clinical implications of neuroscience research in PTSD". In *Annals of the New York Academy of Sciences* 1071(1): *Psychobiology of Posttraumatic Stress Disorder: A Decade of Progress* (July 2006). Van der Kolk, B.A., "The neurobiology of childhood trauma and abuse". *Child and Adolescent Psychiatric Clinics*, 12(2) (2003), 293–317. https://doi.org/10.1016/s1056-4993(03)00003-8. Van der Kolk, B.A., "Posttraumatic therapy in the age of neuroscience". *Psychoanalytic Dialogues*, 12(3) (2002), 381–92. https://doi.org/10.1080/10481881209348674.

4 Changing the system - one process at a time

So far, we have looked at the background to neurodiversity and what your existing perspectives and biases may be around the topic. We have also considered where bias may be institutionalised. In this chapter, we will see how institutionalised bias is present in many workplace systems and processes, such as recruitment and talent development. We invite managers and leaders to follow us through the "employee lifecycle", and reflect on what positive changes they might make to their existing practice to improve inclusivity. We will also briefly look at how the physical workspace can be made more neuroinclusive, focusing in particular on those areas that most managers and leaders have influence over.

A neuroinclusive employee lifecycle

There are five specific stages of the employee lifecycle, as laid out in Table 4.1. This chapter will take each of these in turn and look at how common practice compares against best practice.

Table 4.1 Stages of the employee lifecycle

Attract	• Advertising and job specifications
Recruit	• Assessment process, e.g. interviews
Onboard and integrate	• Cultural adaptation • Role and responsibilities
Manage	• Support and career development • Enabling strong performance
Offboard	• Exit processes

DOI: 10.4324/9781003645535-4

Attract: advertising positions and job specifications

Exercise: neuroinclusive job adverts

Read the two job specification examples and, having reflected on them, answer the following questions:

Box 4.1: Job specification example 1

Head of Communications and Marketing

Serve as a key leadership team member, responsible for coordinating company priorities and projects. Duties include leading staff units in the C-level office, developing initiatives, managing special projects and crises, collaborating with senior officers, communications planning and stakeholder engagement.

As a crucial part of the leadership team, this role requires overseeing various strategic functions within the organisation. This includes setting objectives, aligning departmental goals with the company's vision and ensuring efficient execution of tasks. The individual must possess strong leadership skills to guide and mentor staff members, fostering a collaborative and productive work environment.

In terms of project management, the responsibilities include identifying potential risks, creating contingency plans and ensuring timely delivery of project milestones. Crisis management involves quick decision-making and effective problem-solving to mitigate any adverse impacts on the organisation.

Collaboration with senior officers is an essential part of the role, resulting in harmonisation of cross-functional efforts. This role also entails crafting comprehensive communication strategies to maintain transparent and consistent messaging both internally and externally.

Engaging with stakeholders is another critical aspect, requiring the ability to build and sustain positive relationships with clients, partners and other key entities. By doing so, the organisation can enhance its reputation, foster trust and secure long-term success.

Salary on application. We operate a 5 days in the office policy.

Box 4.2: Job specification example 2

Head of Communications and Marketing (full-time, flexi hours)

Salary: £40,000-£60,000 annually. Benefits include:

- 33 days annual leave including bank holidays
- No compulsory leave
- Company pension contribution 8%
- Company car allowance
- 6 months full pay paternity, maternity, shared parental leave
- Employee Assistance Programme and private healthcare

Location: minimum 2 days a week in HQ Edinburgh, 2 days home working and 1 day other locations visiting other offices. The HQ in Edinburgh is an open-plan office with ~50 people, including a quiet working space.

The company consists of 98 members of staff with annual turnover of £1 million. It is looking for head of Communications and Marketing after the retirement of the previous job holder. This senior leader will report to the Managing Director and be part of Senior Leadership Team consisting of 4 other heads of department covering Finance, Sales and Marketing, Operations, HR and Sustainability to develop and implement the Communications and Marketing strategy.

Person requirements

The successful candidate must have:

- Minimum of 5 years' experience delivering comms and or marketing projects
- Ability to draft and disseminate press briefing at fast pace during high-pressure situations
- Ability to write and communicate sales and marketing strategy with the senior leaderships team
- The ability to influence and persuade peers and seniors to your way of thinking

- It would be nice if the successful candidate had:
- Previous experience managing a team
- Previous experience working in a fast-paced, high-growth environment
- Previous experience managing projects

We do not expect you to meet all the "nice-to-have" criteria, but you must meet the criteria marked as "must have".

We value diversity and therefore ask for any adjustments to the interview and recruitment process as well as day-to-day working. If you wish to communicate your application in a voice recording, please mail us for details as to how to do so.After a human CV sift, first-round assessments will be held weeks of April 12th and May 19th. These will include 1:1 interviews and practical problem-solving scenarios. Materials for these sessions will be shared in advance.

Please contact Head of Human Resources Dave Howard for further details and requirements. Watch our video here of Dave talking about the role.

Which job specification do you think more clearly articulates the role and its responsibilities?

....

....

....

Which job specification is the more neuroinclusive? Which words or sentences indicate this? (Identify at least five ways in which the two specifications differ.)

....

....

....

Which one most closely resembles the kind of job advert *your* company would put out?

....

....

....

What do you notice about how to make job adverts more neuroinclusive? How might you apply that to your own role specifications?

....

....

....

Box 4.3: How did we get here?

Most job descriptions ask the applicant to have a broad range of skills, qualifications and experience; it is normally such a broad range that the list is meaningless!

Job specifications were originally introduced over a century ago by Frederick Taylor following his observations of factories. The work was much simpler to describe in those days; also, newspapers charged by the line for job adverts, which was a further motive for keeping job descriptions short.

With modern online job boards charging a flat fee regardless of length, that constraint has disappeared and job descriptions have grown. They are often vague, too, aiming to cover every possible base in case the role changes over time. This is unhelpful for ND individuals, who may have significant strengths in specific areas but face challenges with other competencies on the list.

As the above exercise might have revealed, a neuroinclusive job posting can benefit every potential applicant, regardless of neurological type, by creating clarity about the expectations of the role. To help make *your* job postings and applications accessible to a more diverse range of talent, here are our top three recommendations:

1. Make the person requirement clear, specific and concise.
    a. Be specific about the role's tasks and behaviours. There is a tendency for job descriptions to be reused, which makes for vague, generalised language. For example, "assist the senior manager" could cover any number of requirements. Instead, try (for example) "assist the senior manager with reviewing monthly budget forecasts for accuracy and completeness".
    b. Before you assemble a list of skills for the role, challenge yourself to find a balance between generalist skills and ones specific to the role. For example, "excellent communication skills" could discourage people who are dyslexic from applying, as they may assume that writing is a key feature of the job when this might not be the case.
    c. Remove unnecessary qualifications from the job specification. Ask yourself, for example, whether a degree is actually needed, or if equivalent experience would suffice. The education system favours some groups over others (e.g. those with dyslexia[1] are disadvantaged), so an unnecessary qualification requirement can introduce unnecessary bias.
    d. Avoid vague phrases such as "team player", "friendly" or "hard-working".
    e. Make it clear what are "nice to have" (desirable) and what are "must have" (essential) experience and skills: those with more logical brains may see a long list and take them all as necessary requirements and therefore not apply.
    f. Include a description of the working environment, e.g. "open-plan office of 200 people". Those who don't thrive in such environments can then choose not to apply or to request adaptations like access to a quiet space.
    g. Use plain English (see Chapter 5 for more on this).

2. Demonstrate that you consider diversity and inclusion.
    a. Include a DEI (diversity, equity and inclusion) statement that specifically includes neurodiversity and offers accommodations to candidates through the process.

   b. Ask candidates what accommodations they need and give them a list of options, e.g. extra time to complete tasks or respond to interview questions; different questions in different formats to address learning styles.
   c. Itemise the types of flexible working available, e.g. core hours, part-time, annualised hours, working from home, job share.
3. Make application forms as short as possible.
   a. Be explicit about how much information is required in each section, i.e. number of words.
   b. Make it easier to complete the application form by breaking down longer questions into separate steps.
   c. Allow applicants to review and edit their applications multiple times before submission.

Apply these three recommendations to roles at all levels, including senior leadership.

Box 4.4: A note on AI (artificial intelligence)

There has been a lot of discussion in the last few years on whether using artificial intelligence (AI) to review CVs at the filtering stage of a recruitment process reinforces bias or reduces it. However, studies have returned inconsistent results: some suggest that AI increases bias, others suggest that it removes it. For now, a combination of AI alongside a human review is thought to be the most effective and inclusive way of identifying the most appropriate candidates to be interviewed.

Source: Bertail, P., Bounie, D., Clémençon, S., and Waelbroeck. P., "Algorithmes: Biais, discrimination et équité". *HR Today*, 58 (2020). https://hal.science/hal-02077745

Once you have drafted your job specification, we suggest you ask your ND community members to review it: they may give you further pointers on making it more inclusive.

When you advertise the role, ensure that you are reaching as broad a range of participants as possible. Partner with community groups and charities who work in the neurodiversity space to avoid ruling out valuable ND talent.

Recruit: assessment processes

Exercise: reflection on interviews

Take yourself back to the last time you were being interviewed for a role that you really wanted. How much clarity did you have about what the role required? Were you provided with high-level, general information or detailed role specifics? If you managed to discover more details, how did you do so?

....

....

....

....

How much advance information did you have about the interview process? For example, did you know who you were going to meet, what they were looking for, the interview format and length, how they would choose the successful candidate?

....

....

....

....

How did you feel going into the interview? What influenced those feelings?

....

....

....

....

Now put yourself in the place of this person:

> You saw a role that you thought you could do: in fact, it was the role you had been looking for. You have the skills and the

experience they asked for. However, you've had bad some experiences in interviews. You find the process stressful, making your brain race around, and you lose focus. You feel anxious when they ask you questions. According to interview feedback, you hurry your answers, but that's because you don't like everyone watching you: you find it difficult to be the centre of attention. You have little understanding about this upcoming interview process and this uncertainty makes you feel even more uncomfortable. You've been wondering whether to disclose your neurodivergent conditions but you don't want to put them off. You don't want to make a fuss or be treated differently. You also worry that they might make assumptions about you based on your neurodivergence, which has happened before. They are bound to have lots of applicants. This seems like a lot of stress and the job isn't even a promotion. You wonder if it's worth it.

What emotions or feelings might you have if you were this person?

....

....

....

....

What impact do you think that might have on your ability to prepare and perform in an interview?

....

....

....

....

How might this impact the effectiveness of the company's hiring process?

....

....

....

....

Assessment processes

The ideal outcome of any recruitment process is that you find the best candidate for the role, someone who will perform well in regular day-to-day conditions. As such, the assessment process should – in form and style – match the organisation's day-to-day conditions, as far as that is possible. Broadly, this means that candidates should be made as comfortable as they would if they were already in the post.

Sadly, this is often not the case. Surveys have found that one in five neurodivergent applicants said that they had been laughed at during interviews;[2] and only 16% believed the recruitment process was fair.[3]

To help address shortcomings like these, it is useful to reflect on these possible interview scenarios. Give your responses before moving on to the following section in which we propose some best practice.

Table 4.2 Suggested changes to the assessment process

What happens	*How might that be received?*	*Try this instead*
The interviewer asks hypothetical questions.	The candidate is required to imagine a hypothetical context and a solution. For some, imagining in this way is very difficult, and can take up a lot of time and energy.	Ask the candidate to recall facts, i.e. what they did in the past and what they learnt from that situation. Studies have shown that structured interviews – where candidates are asked a set of predetermined questions – are more effective at assessing the capability of candidates than free-flowing conversations.[4] So share lists of questions in advance and stick to the script, except when you are clarifying or building on their answers. For an example of this, look to the retailer John Lewis, who share interview questions on their website for the benefit of anyone coming to meet them.[5]
The interviewer asks the candidate for a list of strengths.	The candidate is required to list their strengths. ND individuals can find positive reflection on their qualities difficult.	
The interviewer asks questions that require "crafting" a narrative, e.g. What is your biggest development area?	Those who have a logical and literal approach to questions can find this difficult to answer without creating a negative perception. For example, a neurodivergent brain may choose to share their first, or most accurate, response with all the details, such as saying that their biggest development area is arriving at work on time.	

(Continued)

Table 4.2 (Continued)

What happens	*How might that be received?*	*Try this instead*
A traditional interview format, with questions asked about previous experiences and expertise.	The interview process assesses the social skills of the candidate, not their ability to do the actual job. For example, if the job is to solve problems in electronic equipment, the interview will only test their ability to talk about fixing the machines, not their ability to actually do the fixing.	Offer different types of assessment to accommodate how an individual might best present themselves, e.g. practical assessments, writing exercises, case study analysis, hands-on assessment. EY, for example,[6] have replaced interviews with applied tasks that assess problem-solving capabilities like using Lego.[7††] Flexibility in assessment approach is important, as approximately a quarter of all people find group discussions and pre-prepared presentations difficult. Consider whether timed tests are really necessary considering the role. Such tests might just evaluate processing speed, rather than skills or knowledge. Wherever possible, allow candidates as much time as they need to answer questions or complete tasks.
Interview questions are asked in real time.	This can amount to a test of whether a candidate can deliver a verbal response sufficiently quickly. This disadvantages those who need additional processing time, particularly when under pressure.	Provide the questions before the interview in order to reduce anxiety and eliminate pressure. Allow time for breaks during interview processes that run for several hours.
The environment within which the assessment is held has many sensory stimuli, e.g. room noise, video calls and lighting.	This can become overwhelming for a participant who is processing the stimuli and not able to pay attention to the interviewer.	Check that rooms are neuroinclusive, e.g. where possible they use natural lighting and eliminate background noise. Ask candidates what they need to feel comfortable and perform at their best. If it's a virtual assessment or interview, offer a range of formats, e.g. video call, telephone call.

(Continued)

Table 4.2 (Continued)

What happens	*How might that be received?*	*Try this instead*
Lengthy and multi-step instructions for the assessment are provided in only one format.	The amount of information can be overwhelming for a participant and for some may become a test of memory. Some people may find certain formats (e.g. written) difficult to engage with.	Provide the instructions in multiple media. Give candidates time alone to read instructions or watch a video that explains the process. Avoid black ink on white background for paper and presentations; instead, use dyslexia-friendly colour schemes, e.g. pink, green and blue backgrounds or paper.
Apps and technologies are used which may not be familiar.	The candidate feels uncertain about using the technology, adding to the anxiety about the process.	Provide instructions about platforms or technology being used within the interview process. Ensure tests are compatible with assistive technology and software that candidates might use, such as voice-to-text.
Accommodations or adjustments are offered in the job interview but not in the actual job (or vice versa).	A false expectation is created about what kind of support might be offered in the job. A further challenge is that accommodations are frequently not offered proactively. Estimates suggest that just 17% of companies do this, while it is estimated that two-thirds do not provide accommodations even when requested.[‡]	Be proactive in offering adjustments and accommodations in the recruitment processes that will be available in the actual job. Best practice is to also offer these for all individuals as they prepare their CVs and cover letters, as this demonstrates the openness to make adjustments, e.g. providing voice to text software to prospective job candidates.
Limited information is provided on the interview process, e.g. logistics, timings, what will happen, who is involved.	Uncertainty about what will happen on the day can increase anxiety levels.	Provide clear and specific information ahead of time, including timetable, identities of the people being met, areas of focus for each stage in the process and what will happen after the interview/assessment, including a time-frame. Provide this information in bullet points if possible.

(Continued)

Table 4.2 (Continued)

What happens	*How might that be received?*	*Try this instead*
The interviewer asks multiple questions in one go, e.g. "Tell me about a time when you were asked to solve a problem. What did you do? What was the outcome? What did you learn from it?"	This can be a test of memory rather than skills: the candidate may become muddled or feel overwhelmed. Of ND individuals, 37% have panicked in an interview when the question structure was overly complicated.‡	Provide simplified questions ahead of time and allow candidates to bring notes to the interview. Ask each part of the question separately. Give the candidate time to write notes and then answer verbally if they require it.
Face-to-face interviews often rely on quickly building rapport with new people.	Some ND people find it difficult to decipher social cues, e.g. when to answer, and this can impact likeability and rapport.	Try to eliminate liking and confirmation bias from your assessment process by challenging yourself to focus on what the candidate said and only those behaviours that are relevant to the job.

What would you do if a candidate...

a) ...discloses in the interview that they are neurodivergent?

....

....

....

b) ...is flapping their arms repeatedly during the interview process?

....

....

....

c) ... asks for extra time to perform a task, but the amount of time they have asked for is longer than they would have if they were actually doing the job?

....

....

....

d) ... completes a paper-based task in the required time but it has lots of spelling mistakes?

....

....

....

e) ... has a less traditional employment history on their CV? They have moved around between numerous jobs. However, they performed well in the assessment.

....

....

....

f) ... performs well in the assessment but doesn't make eye contact with you in the interview?

....

....

....

g) ... provides lengthy answers to your first few questions? At that rate, they will not have sufficient time to get through all the questions you need to ask.

....

....

....

h) ...takes a long pause before answering the interview questions?

....

....

....

IDEAL ANSWERS TO THE ABOVE EXERCISE

a) Thank them for disclosing. Ask what needs and adaptations they might require. Provide a list of available choices as a starting point.
b) As with any repetitive behaviour, unless it's causing disruption for others, continue the interview. It might be that this behaviour helps them settle, or it may be a sign of stress. At an appropriate time, ask if there is anything you can do to make the interview more comfortable.
c) Give them the time requested and discuss the outcome, e.g. "Usually we would expect this to be complete in *x* [time]". Challenge yourself to consider whether it would be a problem if the task took longer day-to-day providing it was done to a high standard. Ask if there are any accommodations not already provided that might make a difference.
d) Ask them if they have considered using assistive technology to correct spelling. Is this something they have had feedback on before? How have they approached it previously?
e) There are many reasons for a non-traditional career path: for example, a continuous desire to learn new things. Or maybe the candidate has experienced poor management or discrimination in previous roles. Rather than judging based on their past history, rely on the assessments that are part of the process, as well as personal references. It is also worth asking what they would need a company to provide in order for them to remain longer-term.
f) Does the role require eye contact? For example, is it a face-to-face customer service role, or is it an internal role with mainly email and team-level communication? If eye contact is not a job requirement and the task has been completed, it doesn't need to be raised.
g) If a candidate answers in great detail, ask to them to pause and let them know they've answered the question. It can be helpful to include this in the interview instructions - i.e. the fact that the interviewers may do this - so that the interviewee is prepared if it happens.

h) Wait patiently for the answer, in order to minimise pressure and allow for additional processing time. If the candidate appears uncomfortable or distressed, ask if there is anything you can do to help support them. They may want a few minutes to make some notes and think before answering out loud.

Box 4.5: Auticon's assessment approach

Auticon is a company that hires exclusively autistic IT developers who build IT solutions that enable other companies to become more neuroinclusive. They have a clearly laid-out interview process which involves:

1. A short (1–2 sides of A4) application form
2. An informal interview that is focused on the candidate's passions and what they need to thrive at work
3. Completion of a technical survey
4. Completion of a skills assessment

If a candidate passes these stages, they are placed within their talent pool and, when a job comes up, assigned to a role that suits their strengths. In this way, they focus the assessment process on the candidates' skills and capabilities as they relate to the day-to-day role.

Source: Auticon, "Careers". Auticon, 2025. Available at: https://auticon.com/uk/careers/

Have you ever heard something like this?

Reflect on the statements in Table 4.3 – all of them commonly heard. What misconceptions or assumptions are being made by the person making the statement?

Culture fit versus culture add

Companies frequently talk about candidates being a "good culture fit". This approach encourages people to hire "people like them" and

Table 4.3 Statements and assumptions made by interviewers about ND interviewees

Statement	*Assumption being made or misconception being applied*
"His facial expressions didn't convey enthusiasm for the role."	
"They failed to establish normal eye contact and appear to be antisocial. This would not be good for business."	
"When I asked him what his biggest development area was, he told me that it was turning up to work on time."	
"Her rocking back and forward during the interview concerned me. It seemed like she cannot control her behaviour."	
"I do not want to progress him to the next round as the office is a loud place and he doesn't seem able to handle the noise."	
"They didn't understand my sarcasm; I don't think they'd be a good culture fit."	
"Their answers were great but they didn't stop talking. Did you notice how they talked over questions? We need someone who has better listening skills."	
"If they cannot turn up on time at the right place they won't be reliable enough to work for us."	
"Though I liked their answer, and it could be feasible, it's not the answer we had thought of."	
"They seemed like hard work - not understanding the questions and asking me so many questions. They also seemed to disagree and ask 'why?' a lot. A troublemaker in the making."	
"They have the right skills on paper but they seem set in their ways with a fixed mindset."	
"When they gave that example of how they don't like to be told what to do I thought they would be difficult to manage."	
"When I asked 'Where do you see yourself longer-term?' and they said 'as MD'. Who do they think they are? They must have a big ego."	

exclude those who differ. Studies have shown that job candidates who are similar to the interviewer, in terms of world-views, linguistics and background, are favoured in interview situations.[8] Instances like this are examples of liking bias, which we covered in Chapter 2. It's easy to see how such an approach could disfavour someone who has neurodivergent traits or perspectives, and likely differing from their interviewer.

Instead of thinking about whether someone might "culturally fit" within the organisation, managers might consider what they might *add* to the culture. This would mean purposively hiring people who are different in some way to others in your team, albeit still in line with the values of the organisation. There may also be commercial benefits: for example, hiring someone with a different perspective could help you connect to customer groups with the same perspective.

Disclosure during the interview process

Not all neurodivergent individuals have been diagnosed; even if they have been formally diagnosed, they may not want, or believe they need, to disclose it to an interviewer. Studies back this up:

- Disclosure of autism has been linked to a ~26% reduction in expressions of interest from prospective employers versus those who did not disclose a condition[9]
- One in six ND individuals have had job offers rescinded because of their neurodivergent conditions[10]
- Two-thirds say employers see ND as a red flag[11]

Disclosure during the recruitment process can:

- Increase protection against blatant discrimination
- Provide an opportunity to discuss needs and assess accommodations
- Provide a supportive environment
- Increase the chances of maintaining potential employment once they have disclosed

On the other hand, disclosure can:

- Generate stigma and subtle discrimination
- Introduce privacy concerns around medical information
- Generate uncertainty around how information might be received
- Lower the likelihood of advancing to the next stage
- Create envy and resentment from others around accommodations

A decision to disclose therefore involves weighing up different considerations. The best you can do as a manager is make the environment as supportive and inclusive as possible by actively demonstrating support for neurodiversity through your job advert, your words and your actions.

Onboard and integrate

What new recruits need to know

One of the biggest areas of uncertainty for any new starter is figuring out "how they get things done around here". This is seldom defined explicitly, and this lack of clarity can be difficult for neurodivergent individuals to work with. For example, it may be that tea and coffee are made on an informal rota basis, or people are expected to eat their lunch at their desks. Assigning a buddy or mentor to a new recruit can provide a sounding board, someone to clarify these unwritten rules.

Below are examples of the support a company can offer to settle new recruits quickly and smoothly, with the least disruption for both parties.

Before they join:

- Provide tours of the physical premises, including staffrooms, toilets, car parks and kitchens. Reduce anxiety for those who can't get to the physical premises by making a video recording of a tour
- Provide a site map
- Arrange for their technology to be set up before they join, so they can start doing their job as soon as possible. This can help reduce the uncertainty that comes with a new role

On day 1:

- Familiarise the new recruit with the working environment, incorporating an induction into the technology (including assistive technology)
- Provide a written induction pack with key policies, procedures, contact details and online links
- Provide an opportunity to discuss daily routines and an opportunity to shadow a more senior member of the team before the employee is asked to work independently
- Agree expectations and provide clear instructions on tasks for the first week. Defining when and how work will be reviewed and how feedback will be provided can help reduce anxiety and ensure expectations are explicit
- Debrief at the end of the first day 1:1 to answer any questions the recruit has that they might not wish to raise in a group forum or open-plan office
- Make a point of asking about adjustments or what productivity tools would help them do the best job
- Request their neurodiversity passport or adjustment plan, if they have one Neurodiversity passports could contain information about a person's protected characteristics (i.e. diagnosis or disability), their needs and what adjustments have worked for them in the past. It means that they do not have to explain everything from the start and can rely on a written document to provide the information (see Figure 4.1)

The first week:

- Proactively check in and ask how they are doing
- Provide opportunities for employees to socialise with others, so they build up support networks
- Ask and listen to what they say about their preferred working environment and adjust appropriately. For example, you might ask:
    - What do you need (e.g. accommodations, adjustments, ways of working) that might make work easier for you?

- Are there specific environments or settings that work best for you at certain times?
- Do you have a preferred way of communicating or receiving information, e.g. email, text, face-to-face, audio, visual aids?
- How do you prefer to be managed, e.g. frequency of 1:1s/check-ins?

Individual Adjustments Plan

Guidance

If you are completing an Individual Adjustments Plan to accommodate disability needs, please refer to the Reasonable Adjustments Policy – Managers Guidance.

The plan can be used for a wide range of other circumstances and is a record of any requirements employees have, and adjustments which would help them perform their job to the best of their abilities.

The completed form should be saved and stored securely. You will keep a confidential copy of this record and share a copy with the employee. You will be responsible for reviewing the agreed arrangements periodically with the employee to address any issues or concerns which may arise or change over time.

If the employee changes roles or line manager, they will be responsible for sharing this document with the new line manager and should rediscuss the plan with them.

The following checklist can help you to identify, agree and implement adjustments:

Checklist:	**Completed (Y/N):**
Before the meeting to discuss adjustments:	
Familiarise yourself with the relevant policies or guidance	
Read through any relevant information provided by the employee and/or support services	
After the meeting to discuss reasonable adjustments:	
Reflect on support requirements – think of the job and the tasks undertaken by the employee and reflect on the challenges they are facing in the workplace	
Reflect on individual adjustments – consider the potential adjustments and reflect on how they remove or reduce the challenges the employee is facing at work	
Complete all the sections of the Individual Adjustments Plan	
Set a review date with the employee	
Share the Individual Adjustments Plan with the employee	
Safely and confidentially store the document	

Figure 4.1 Example of an adjustment plan or employee passport

Section 1: Employee's details	
Please complete all sections below:	
Name:	...
Assignment Number:	...
College/Professional Service Group:	...
School/Department:	...
Current Job Title:	...

Section 2: Agreed adjustments	
Support Requirements	**Agreed adjustments**

Section 3: Communications and changes
Information about the employee's circumstances will not be shared with third parties without their consent unless required by the law. The Individual Adjustments Plan will be kept securely and confidentially by you and your employee and can be shared with other colleagues at the employee's discretion only. The Individual Adjustments Plan may need to be shared with colleagues whose role is to provide advice on and/or implement the agreed adjustments. If the employee changes roles or line manager, they will be responsible for sharing this document with the new line manager and should rediscuss the plan with them.

Section 4: Conclusion and review	
The adjustments included in this form should be regularly reviewed. Reviews should be conducted at least annually, but they could be more frequent if required, and depending on individual needs. This is to ensure that individual adjustments reduce or remove any identified challenges and that they remain effective at all times. Further reviews can be requested by employees who experience sudden changes in their circumstances that may have an impact on their ways of working.	
Current date:	
Review date:	
Any additional comments:	

Figure 4.1 (Continued)

The first three months:

- Introduce them to groups supporting those with protected characteristics (e.g. employee resource groups, occupational health), if they want
- Hold a review of objectives in order to obtain feedback on what they have experienced and what changes could be made. It is worth asking about what they are enjoying and how they could use their strengths more, so that you can get the most out of them. (See Chapter 7 for more information on strengths)
- Arrange a time to check in with human resources and occupational health about flexible working, equipment for working from home, and other policies

> **Box 4.6: Employee resource groups**
>
> Employee resource groups (ERGs) or support circles, which are voluntary and employee-led, can provide a way for neurodivergent individuals to meet other neurodivergent individuals in other teams, helping to create both a network and a group to provide support. ERGs can also be an effective means of obtaining feedback from employees about what the organisation is doing well and what it could do better. They can provide peer-to-peer support around neurodiversity-related challenges; they might also include neurotypical individuals who are there to learn and support.

Manage

Support

As a manager and leader, you are responsible for ensuring that your team members have the right practical support to perform at their best. The best way to identify what kind of support your team members need is to have an open and honest conversation with them about how and when they work at their best. We have included some tips on how to have those conversations in Chapter 5. Below are some examples of what you can do to meet these requests, although what is possible to deliver will depend on what your company is willing to provide.

Sensory stimulation and workplace setup

- Create differentiated working zones, such as one for "creative" group work and another for "quiet" work. Move noisier groups to one end of the office where possible
- Eliminate background noise by, for example, allowing headphones or removing background music
- Change lights to make them less stimulating, e.g. from strip fluorescent to incandescent lighting
- Be mindful of areas of the office that can be noisy or smelly (e.g. the kitchen). Use odour-free cleaning products, as some individuals are more sensitive to smells
- Be aware of visual stimuli such as clutter, or noticeboards with too much information. Remove unnecessary stimuli where possible
- Turn off office equipment when it's not in use
- Allow use of tinted lenses or glasses to reduce glare and visual sensitivities
- Consider universal design principles, i.e. designing everything for accessibility, usability and inclusivity for all
- Allow for sit/stand desks, so that individuals can move around
- Provide acoustic barriers between desks
- If requested, allow individuals to have a fixed desk rather than hot-desking

Ways of working

- Allow different communication media and accept cameras being switched off in meetings. Provide assistive technology where required, such as text-to-speech software
- Offer written instructions instead of verbal ones and provide structured feedback.
- Agree ways of working around interruptions, e.g. no interruptions for the first two hours of each day
- Encourage flexible working to accommodate preferred working schedules and commutes, with set times to come together as a team
- Provide coaching for neurodivergent individuals in order to help

them to adapt to the workplace
- Provide support and training around common challenges, e.g. time management
- Offer working-from-home and flexible hours

The above list is not exhaustive, and every individual is different, so the key point here is to ask what works best. Some people may not know what works for them, in which case your role as a manager is to partner with them to experiment with some options until you find one that works. Managers should make a point of having regular discussions with their team members about the support they need, as it may change with time. Questions in such discussions might include:

- How do you prefer to build working relationships?
- What kind of recognition resonates with you, e.g. public acknowledgement, private feedback, tangible rewards?
- How do you prefer to receive feedback, e.g. frequency, verbally versus written?
- How do you learn best, e.g. face-to-face, written?
- How can I support your career growth?
- How can I best support your wellbeing?
- When you are at your best, what is your environment like?
- How do you prefer to receive instructions?
- How can I best delegate to you?
- How do you prefer to communicate with others, e.g. video, call, text?
- Do you have any sensory sensitivities that we should be aware of?

Example: GCHQ

Government Communications Headquarters (GCHQ) [12] takes a number of steps to ensure that a diverse range of thinkers apply for roles and grow their careers with them. For example:

- On job advertisements, they endorse the value associated with divergent thinking by stating "we value how you solve the problems in your unique way"

- They have created a Disability and Neurodiversity (DAN) team, which provides comprehensive guidance on how to tackle challenges that can be faced, regardless of whether someone has a diagnosis
- If support is required, they provide access to a specialist neurodiversity advisor: an external contact who provides advice on overcoming difficulties
- They provide affinity groups that create an ND community and mutual support

Career development

Below are some self-reflection questions about career development.

For you, what does a "successful" career path look like?

....

....

....

What do you think of individuals who don't achieve your definition of success?

....

....

....

How do you perceive individuals who move quickly from role to role or who have come into the role from a different route?

....

....

....

What assumptions do you make about how quickly and readily people will adapt to a new role? What is that based on?

....

....

....

For neurodivergent individuals it can be helpful to nuance career development approaches in order to make sure that they fit with the individual's needs and ambitions. For example:

- Some may find it difficult to visualise or imagine themselves in roles other than the one they are in. Providing such individuals with the opportunity to trial a job for a week, or to shadow someone doing that role, can help smooth the transition
- Taking on more work or responsibility can seem overwhelming. Breaking down the steps towards a career, role or responsibility change can help reduce overwhelm
- Create a specialist career pathway for those with specific talents. Make it OK to be a true technical expert. Some may have a narrow set of interests, and staying in their role or the same team may be a preferred career path. Not everyone needs to move around or manage others to progress
- Ensure that promotions and role changes of neurodivergent individuals in particular are communicated broadly across the organisation, e.g. short films about employees and their career journeys. This may inspire others, including those who haven't disclosed, to change roles or progress
- Be aware that change, including transitions between roles, can be particularly stressful for those with autism. There is more on handling transitions in Chapter 8
- Rather than seeing a role move as a way of closing out a development area, consider instead how it might help a person to use their strengths more. We will talk more about using a strengths-based approach in Chapter 7
- Employees with ADHD may be actively seeking change even when they are relatively new in the role, in order to maintain a level of interest. How might a role be enriched on an ongoing basis to enable them to maintain interest?

Neuroinclusive networking and events

Careers are often accelerated and improved by networking. However, the traditional forums for networking - large-group social events in which people mingle and chat informally - can be uncomfortable or daunting for ND individuals (and many neurotypicals, too!). It is worth considering other options to support networking. For example:

- Small-group meetings
- Structured or interest-focused forums
- Virtual networking, allowing for 1:1 discussions

For events more generally, of any kind, it can be helpful to:

- Share detailed pre-event information including timings, attendees, agenda, parking information, photos of the event space, etc.
- Take into account sensory considerations, e.g. flashing lights, noise levels
- Provide communication cards, which can indicate whether people want to talk to others or not. These can be made visible on lanyards
- If questions are being asked (e.g. of an expert panel), allow people to submit them in advance
- Use prompt cards or icebreaker questions to instigate informal, rapport-building conversations. For example, "Tell me about a place you would like to visit - real or imaginary"

Case study: Missed Opportunities

> I think that if I were able to be open about how my mind works, with support, I could probably find a way to work at a higher level, as I have much to offer, but I tend to get overlooked. I am also not good at "putting myself out there".
>
> 58-year-old male in the healthcare sector with AuDHD

This statement reflects how potential can be wasted as a result of lack of support from managers.

Enabling strong performance

Neurodiversity adds an additional consideration to any performance management situation. The keys to success in this space are clarity and curiosity: clarity provided to the employee about both expectations and performance; and curiosity shown by the manager about what enables them to perform certain tasks well and about which tasks are challenging.

Goal-setting

Critical to enabling strong performance is agreement and clarity around goals and expectations. Goal-setting provides direction and focus to team members and can help generate motivation, whereas setting expectations helps reduce uncertainty. However, goals aren't always formulated as effectively as they could be – being vague, for example, or quickly going out of date.

Here are some self-reflection questions on goal-setting:

How were your skills development and task-related goals set? (E.g. through discussion with your manager?) What level of influence did you have in the process?

....

....

....

How explicit and specific are the goals? (E.g. are they quantifiable?)

....

....

....

What will be different when you have achieved those goals?

....

....

....

At that point, will there be any room for discussion about whether they have in fact been achieved? (I.e. could there be lack of clarity or room for interpretation?)

....

....

....

Which parts of the outcome were explicitly defined by your manager and which parts were inferred by you?

....

....

....

Is there any difference in your level of motivation or engagement around goals that are clearly and explicitly defined compared with those that are more vague?

....

....

....

Here are some recommendations for enabling strong performance among your neurodivergent team members:

- Understanding what skills, behaviours and experience are judged as being "successful"
    - Be explicit and objective about what is required to progress, e.g. "To get promoted you need to have at least six months' experience in managing people..."
    - Document a plan around skills development that is frequently maintained and reviewed between both of you
    - Clarify the importance of extracurricular activities, e.g. managing team social events, attending networking events
- Goal-setting
    - Enable choice in the goal-setting process: this increases motivation and engagement. Be aware that, for some, the decision-

making process is easier when the options are closed (i.e. "Would you like *this* or *that*?") rather than open-ended (e.g. "What goal would you like to achieve?")
  - Give the employee time to review performance goals prior to meeting with their manager
  - Be flexible about how outcomes may be achieved so the employee can be rewarded by finding creative ways to reach them
  - Clearly define performance goals, in writing and using visual practical examples, specifying the important dimensions of performance and giving a clear definition of what the completion of the goal looks like. Offer to break goals down into manageable steps if this helps
  - Specify why these goals are important to the organisation as a whole and what difference their contribution can make. Be explicit about the impact if the goals are not achieved, e.g. on client relationships, teams or project timelines
- Measurement of progress
  - Create tools or data sets that allow ongoing evaluation of performance. These can be quantitative or quantitative
  - Talk about wellbeing at the same time as performance and task goals. This demonstrates that you care about the holistic person and not just their work-related outcomes
  - If an employee is not hitting expected performance levels, start by being curious about what lies behind that: ask what additional support they might need, or how they might be able to use their strengths to complete tasks (see Chapter 7 for strengths-based approaches). It is also worth considering what influence they may have on the outcome, i.e. what is within their control.
  - Provide feedback in alternative formats (e.g. written) to complement in-person feedback. More on this below

Typical challenges around feedback

Providing good-quality feedback as close to real time as possible has been shown to enhance motivation and improve performance. The following are some common challenges that neurodivergent (and often neurotypical) people experience around feedback:

1. Vagueness. For example, "This report wasn't very good"; "You need to improve your communication". This does not provide a clear idea of what to do differently.
2. Difficulty understanding another's perspective. For example, an autistic individual providing direct feedback to someone in a meeting about the gaps in that person's report might believe they are simply being honest. However, the report writer may find it rude or disrespectful, and feel hurt. Understanding another's perspective can be difficult for some ND individuals, due to challenges with what is termed "theory of mind".[13]
3. Difficulty expressing needs. Some individuals may encounter difficulties if addressing the feedback requires additional support. If that appears to be the case, give the employee time to process and ask them to come back later with thoughts on how they can best be supported. You might also invite their mentor to the performance review, as a supportive presence and to facilitate a discussion about needs.
4. Feedback that addresses implicit expectations. Feedback can sometimes concern itself with an ND individual not adhering to the office's unwritten code of conduct. Before delivering feedback, check that it relates to expectations that have been explicitly laid out. It may simply be the case that your team member needs to have the expectations set out for them.
5. Feedback relates to the process not the outcome. ND individuals can have different ways of getting tasks done.
6. Feedback that is delivered in real time. Some individuals require time to process feedback before they can have a productive conversation about it. Ask them if they prefer to have verbal or written feedback first, i.e. in what order.

Case study: Disabled Staff Network

The University of Edinburgh's Disabled Staff Network (DSN) has the primary aim of providing a network of mutual support and a collective voice for staff who are disabled and/or neurodivergent. The DSN closely works with Human Resources, in particular with the Equality, Diversity and Inclusion (EDI) team, to progress disability issues and make improvements.

The DSN holds six meetings a year and maintains an email distribution list via which they share and request information around disability and neurodivergence. It is chaired by two senior members of staff, who are supported by two other officers to help oversee communications.

Remit:

- Bring together staff who identify as disabled and/or neurodivergent in a supportive and confidential environment to share experiences, information and support, to discuss disability issues and to network
- Gather and share best practice, knowledge, ideas and thoughts on disability issues to benefit operational, research and academic work
- Inform the university's approach to creating an inclusive, accessible and supportive culture and help to develop effective solutions by acting as a forum and a point of contact for the university to consult with on relevant policies and projects
- Raise the profile of disabled staff and disability within the university and beyond

Current aims:

- Create branding and related advertising resources
- Advertise the network to grow membership, increase volunteer involvement and raise the profile of disability
- Gather information about existing support for disabled and neurodivergent staff within the DSN SharePoint

- Set up a university-wide Neuroinclusion Group on MS Teams
- Enable use of sunflower lanyards, wristbands or pins to signify non-visible disabilities
- Raise awareness of the meaning of sunflower lanyards
- Collaborate directly with network members to hear their opinions and thoughts

Member and EDI representative's experience:

> My experience of the DSN has been overall positive. I joined the network one year ago, not long after I joined the University. As an ADHDer and a person with a long-term health condition, joining the network made me feel supported and able to openly share my challenges and my fears in a non-judgemental and unassuming space. The network is a great space because it gives disabled and/or neurodivergent colleagues the opportunity to share their perspectives on potential improvements that could be made at university or local levels, share best practice and offer creative solutions based on lived experience and, ultimately, share challenging or traumatic experiences with a group of peers that may be able to provide practical support, or give a space to be listened to.
>
> Federico Marchiolli, HR Partner - Equality, Diversity and Inclusion

Offboard

For every step in the employee lifecycle clarity of communication is critical, and no less so when you are looking to offboard an employee, whatever the reason. For many, whether neurodivergent or neurotypical, this period can be very challenging. And it can be so much worse for those who find change difficult or are less likely to get another job because of discrimination. Here are some principles for managers and leaders to consider:

- Communicate clearly and often, allowing time for messages to be digested. Though you may have explained everything the first

time round, multiple conversations may be required to allow for questions and understanding
- Communicate in writing as well as verbally, to ensure the communication has been received and understood
- Provide resources and support to aid the employee in their next career step, e.g. recruiters, career services/coaching and courses
- Provide information about who references can be requested from
- Discuss how they want to signify their exit from the company, e.g. an email to everyone, a leaving celebration, a quiet lunch with immediate co-workers
- Clear details should come from HR about what their last days and weeks will look like. This includes instructions on how to return equipment, where relevant
- Ask what type of exit interview they would like, i.e. in-person or written

The legal requirements around support (UK)

The Equality Act 2010[14] legally protects certain categories of people from discrimination in the workplace, as well as in wider society. It applies to all employers (regardless of size) and to all workers (regardless of contract type) and job applicants.

There are nine "protected characteristics" under the Act, including age, gender reassignment, race and disability. Disability is defined as a physical or mental impairment that has a substantial and long-term adverse effect on an individual's ability to carry out normal day-to-day activities. Neurodivergent conditions may fall within the definition of "disability" for the purposes of the Act and, with that, afford neurodivergent individuals protection against unlawful discrimination.

The Act prohibits several different forms of discrimination:

- Direct - where someone with a protected characteristic is treated less favourably than others because of their protected characteristic. An example is an employer refusing to hire someone because they have autism
- Indirect-where acts, decisions or policies (broadly speaking) which are not intended to treat anyone less favourably have the effect,

in practice, of disadvantaging a group of people with a particular protected characteristic and there is no justification for that disadvantage. For example, if a company requires all employees to work in a noisy, open-plan office, this may disadvantage those with sensory challenges who are sensitive to noise and distractions
- Harassment - where there is unwanted behaviour that is related to a protected characteristic, and the behaviour violates the protected individual's dignity or creates an intimidating, hostile, degrading, humiliating or offensive environment for them. Examples are people mocking repetitive movements that can come with autism, or mocking someone with dyslexia for not being able to spell correctly
- Victimisation - where someone who has complained about discrimination or harassment is treated unfairly because of that
- Discrimination arising from a disability - whereby someone is treated unfavourably because of something arising from their disability. For example, if an employee with ADHD is penalised for taking regular breaks to help maintain their focus level, this could fall under this category of discrimination. It is worth noting that employers are also required to make reasonable adjustments to ensure that disabled applicants and employees are not at a disadvantage in the workplace due to workplace practices or physical features. Reasonable adjustments for neurodivergent individuals could, for example, include providing extra time to complete assessment tests or providing ear defenders if there is noise in the office

The Act applies to the whole lifecycle of the working relationship, i.e. recruitment, dismissal, employment terms and conditions, pay and benefits, promotion and transfer opportunities, training and redundancy.

Public bodies, such as universities, government departments and the National Health Service, may also be impacted by the public-sector equality duty contained in the Act. This imposes additional obligations aimed at eliminating discrimination and advancing equality of opportunity. (See the resources list in Chapter 12 for specific public sector duty.)

Hate crime laws can apply in respect of neurodivergent conditions, too. A hate crime can occur where a victim believes that a crime was carried out because of prejudice towards disabled people. It can be considered a hate crime if people are targeted because they are believed to have a condition, even if they do not actually have it. For example, if someone attacks someone else because they believe that the person has dyspraxia, it can count as a hate crime even if the individual does not have dyspraxia. Hate crimes can be verbal or physical and include physical abuse, threatening behaviour, mocking, teasing, online abuse and damage to property.

There are also general data protection rules around confidentiality and storage of data that is sensitive, such as information about workers' neurodiversity-related diagnoses and treatment.

Asking about neurodivergence

In interview situations, employers ought not to ask job applicants directly about health and disabilities (including neurodiversity). To do so may be unlawful under the Equality Act 2010. Employers can only ask about a disability in an interview if:

1. It is necessary to find out if an applicant will be able to carry out an intrinsic part of the job that cannot be met with reasonable adjustments
2. The employer is finding out if an applicant can take part in a future assessment/interview to test their ability for the job, or if the applicant requires reasonable adjustments to take part in a future assessment/interview
3. The employer is taking positive action in relation to disabled people
4. Having a particular disability is an occupational requirement of the role for which the applicant is applying
5. Another legal requirement requires the employer to ask health- or disability-related questions (e.g. national security)

To be permissible, any question asked about a disability must be necessary for one of the purposes listed above. If an employer asks an applicant about a disability for any *other* purpose and then acts in reliance on the information disclosed (for example, by not awarding the applicant the role), that may amount to a form of unlawful disability discrimination.

Getting financial support from the UK government

In the UK at the time of writing there is a publicly funded employment support programme called Access to Work. It provides practical and financial support to help individuals to overcome barriers to working based on health conditions or disabilities. Funding can be received for areas such as workplace assessments, assistive technology, coaching or support workers and neurodiversity training. No diagnosis is required to access the funding.

Key points from Chapter 4

1. Taking a more inclusive approach across the employee lifecycle - hiring, firing and developing - will benefit everyone. At every stage in the process, ask what the other person needs in order to do their job and how you can support them, as well as being proactive in offering accommodations.
2. Common recruitment practices are based on neurotypical behaviours to recruit people who fit an average profile. To attract and recruit more diverse talent, changes to these practices are required. Such changes often imply more effort but cost nothing and can benefit everyone.
3. Start with a blank sheet of paper when it comes to performance management. Make no assumptions about what the individual's goals are. Provide clarity with regard to expectations and feedback; be curious about the individual's experience when discussing performance.
4. Where performance management is required, focus on clarity and transparency throughout the process in order to reduce the stress and anxiety it can cause.
5. Ensure that you are adhering to or exceeding the requirements laid down by your local lawmakers and policies.

Notes

1. Thompson, A.M., Wood, C., Thompson, I.K. and Carroll, J.M., "Seeing, being seen and being able to see dyslexia in English schools: Parent and teacher perspectives". *Journal of Dyslexia: An International Journal of Research and Practice*, 31(2) (24 March 2025). https://doi.org/10.1002/dys.70003
2. Zurich, "Excluded from the job market: Almost half forced to hide their neurodiversity". Zurich Insurance, based on OnePoll survey data of 1,000 neurodivergent UK adults. Fieldwork carried out 1-25 October 2024. Available at: https://www.zurich.co.uk/media-centre/excluded-from-the-job-market-forced-to-hide-their-neurodiversity
3. Pearn Kandola Research, *Neurodiversity at Work (2024)*. Pearn Kandola, 2024. Available at: https://pearnkandola.com/wp-content/uploads/2024/07/Neurodiversity-At-Work-Report-2024.pdf
4. Rivera, L.A., "Hiring as cultural matching: The case of elite professional service firms". American Sociological Review, 77(6) (2012), 999-1022. https://doi.org/10.1177/0003122412463213. Bergelson, I., Tracy, C. and Takacs, E., "Best practices for reducing bias in the interview process". Current Urology Reports, 23 (2022), 319-25. https://doi.org/10.1007/s11934-022-01116-7
5. Labiak, M., "John Lewis job applicants get questions beforehand". BBC News, 26 April 2024. Available at https://www.bbc.co.uk/news/articles/c6pyqjm1219o
6. Zurich, "Excluded from the job market: Almost half forced to hide their neurodiversity". Zurich Insurance, based on OnePoll survey data of 1,000 neurodivergent UK adults. Fieldwork carried out 18-25 October 2024. Available at https://www.zurich.co.uk/media-centre/excluded-from-the-job-market-forced-to-hide-their-neurodiversity
7. Zurich, "Excluded from the job market: Almost half forced to hide their neurodiversity". Zurich Insurance, based on OnePoll survey data of 1,000 neurodivergent UK adults. Fieldwork carried out 18-25 October 2024. Available at https://www.zurich.co.uk/media-centre/excluded-from-the-job-market-forced-to-hide-their-neurodiversity
8. Ramsahye, Nilma, *Doing Cultural Fit in Job Interviews: The Challenge of Diversity in Recruitment Practices*. PhD thesis, Birkbeck, University of London, 2022.
9. Romualdez, A.M., Heasman, B., Walker, Z., Davies, J., and Remington, A., "'People might understand me better": Diagnostic disclosure experiences of autistic individuals in the workplace". *Autism in Adulthood*, 3(2) (2021): 157-67. https://doi.org/10.1089/aut.2020.0063
10. Zurich, "Excluded from the job market: Almost half forced to hide their neurodiversity". Zurich Insurance, based on OnePoll survey data of 1,000 neurodivergent UK adults. Fieldwork carried out 1-25 October 2024. Available at: https://www.zurich.co.uk/media-centre/excluded-from-the-job-market-forced-to-hide-their-neurodiversity
11. Zurich, "Excluded from the job market: Almost half forced to hide their neurodiversity". Zurich Insurance, based on OnePoll survey data of 1,000 neurodivergent UK adults. Fieldwork carried out 1-25 October 2024. Available at: https://www.zurich.co.uk/media-centre/excluded-from-the-job-market-forced-to-hide-their-neurodiversity

12 Government Communications Headquarters (GCHQ), "Life at GCHQ". UK Government, 2024. Available at: https://www.gchq.gov.uk/section/culture/life-at-gchq. GCHQ, "Diversity and inclusion". UK Government, n.d. Available at: https://www.gchq-careers.co.uk/life-at-gchq/diversity-inclusion.html.
13 Baron-Cohen, S. *Mindblindness: An Essay on Autism and Theory of Mind*. Cambridge, MA: MIT Press, 1995.
14 Government Equalities Office and Equality and Human Rights Commission, "Equality Act 2010". UK Government, 2010. Available at: https://www.legislation.gov.uk/ukpga/2010/15/contents

5 Communication – the key to everything

Introduction

An ability to communicate effectively is critical to every relationship, including between a manager and team members. Effective communication is what helps us to feel connected, to be engaged, to learn and to understand each other. When workplace communication is based on trust, collaboration and respect, it supports wellbeing[1] and enhances productivity.[2] Conversely, when communication is inadequate or breaks down, the result can be misinterpretation, negative judgement, unconstructive conflict and – as we will see in later chapters – potentially even career- and life-changing actions. Neurodivergent individuals are more likely to think and communicate differently to neurotypical norms,[3] which presents increased potential for communication breakdown. It is therefore all the more important to find ways to communicate effectively with this population in order to reduce the risk of such negative outcomes, and to get the best out of your entire team.

In this chapter, we will be looking at communication, and we will examine some approaches that can help you avoid the pitfalls. We will begin with some basic principles around good communication before applying it to specific managerial situations.

Throughout this chapter we will be keeping in mind the "3Cs" that underpin good communication:[4]

1. **Clarity**. The aim of communication is to get the message across. Messages land better when they are straightforward and understandable.

DOI: 10.4324/9781003645535-5

2. **Courtesy**. People are more likely to be open to receiving your message when you use a positive, respectful and considerate tone.
3. **Consistency**. Don't say one thing one day and something else the next. This is important in maintaining trust among your team members.

The impact of communication

Consider a time when you worked with someone whose communication approach was noticeably different to yours. Maybe they replied to emails in a different manner, talked a lot more than you in meetings, used language and phrasing that you weren't familiar with or were more (or less) direct than you were used to. What impact did this have on you and your relationship with that person?

....

....

....

....

How might neurodivergent and neurotypical communication differ?

Neurodivergent individuals can differ from the neurotypical in the ways in which they verbally communicate,[5] most notably in terms of non-verbal cues, verbal rhythm and content. While it can be helpful to reflect on how different neurotypes experience the world, every individual is different so always ask about each person's preferred communication approaches.

Challenges in understanding non-verbal cues are bidirectional: both ND and neurotypical individuals can find it difficult to understand the non-verbal cues of the other type. Some examples are:

1. *A neurodivergent person stares out of the window rather than looking someone in the eyes when they are being spoken to.* Staring out of the window may help them to focus and reduce the discomfort associated with eye contact. However, from a neurotypical

perspective, this can be perceived as not being listened to, or being rude, not caring or untrustworthy.

2. *A neurotypical person communicates their disagreement with a workplace decision using a facial expression rather than words.* This message can be missed or misinterpreted by an ND individual.
3. *A neurotypical person folds their arms midway through a conversation.* Many interpret this (correctly in some cases) as a sign of growing impatience. ND individuals will likely not pick up on this as they generally find it difficult to read body language.

As a manager, there are two key things that you can do to help prevent these kinds of misunderstandings:

1. Identify and challenge your assumptions. For example, when you see the person in example one staring out of the window and find yourself assuming that they are not listening to you, ask them a question (in a respectful tone) to check if that is actually the case.
2. Consider how often you use non-verbal cues to communicate your message. How might you express yourself verbally instead?

Box 5.1: Sensory challenges

Most neurotypical people can filter out irrelevant sensory information in their immediate environment. In a busy restaurant, for example, they can focus on the conversation around the table, ignoring the conversations on other tables, the sounds from the kitchen and the background music. Neurodivergent individuals are often unable to execute this filtering process to the same extent, and sometimes not at all.

As a sense of what that might be like, imagine walking into a meeting. Six people are moving around, chatting about their weekends and laughing loudly. As they begin to settle into their seats, chairs are scraped along the floor into position. The strip lighting is flickering and making an intermittent buzzing noise. The air conditioning is on, humming in the background. You notice the cold it creates on your left arm and the warmth of your right arm where the air flow doesn't reach. The table moves as people pull themselves into their seats. There's a scent of after-shave, stale air and sweat. You notice the pattern on the carpet and discarded tissues on the floor. All this data is coming at you at the same time and at the same intensity: there's no way for you to stop or prioritise what's coming in. As the meeting begins, how might this impact your ability to focus on what's being said?

Communication can also differ with regard to its pace and rhythm. For example:

- Some people may need more time to process incoming data (see Box 5.1). Maybe you will need to become more comfortable with sitting in silence for a little longer than usual while they process. Alternatively, you might need to pause and return to the conversation later
- Individuals with ADHD may talk for longer than might be expected, sometimes going off-topic. As their manager, having identified this as a communication trait of your employee, perhaps agree with them in advance whether you have permission to interrupt if you hear them doing this. If interrupting them verbally is uncomfortable for either of you, agree on non-verbal cues instead, such as holding a hand up to indicate that you would like them to stop talking or using a hand gesture of a circle to ask them to circle back to original topic

The content of communication can differ, too. For example:

- Some neurodivergent individuals may skip small talk in favour of getting onto the topic. Perhaps they feel a lack of certainty about the correct moment for the small talk to finish, and therefore the small talk causes anxiety.[6] This tendency to plunge straight into the topic at hand can come across as rude, despite no such intentions.

In such situations, challenge your own assumptions about what the person meant to convey, if anything, by dispensing with the small talk

- Oversharing can also be a neurodivergent trait. In most workplaces, sharing highly personal stories about yourself is not a social norm. Those with ADHD, or those who find it difficult to identify social norms, may find themselves sharing details of their personal life, even within a short time of meeting someone. Others may perceive this as being a bit odd and find it uncomfortable. If one of your team members does this, and you don't wish them to continue, have a conversation with them about what is appropriate in work situations (see the section below on feedback conversations)
- Some ND individuals can take figurative language literally. As an example, if you were to talk about "letting the cat out of the bag", some people could take this to mean that you had an actual cat in a bag, prompting questions about why you put a cat in a bag in the first place and if the cat was OK. Common phrases like "I'll be back in five minutes" can imply that you will be back in *exactly* five minutes, leading to confusion. Try to be specific, using plain (non-figurative) English (see below in this chapter)

In addition to these examples, other difficulties for neurodivergent individuals include written communication (for those with dyslexia) or controlling communication (for those with Tourette's or ADHD).

Assumptions about communication

Table 5.1 invites you to look at considerations around communicating. In communicating generally, we adapt how and what we say depending on the audience, and this should also apply when communicating with neurodivergent individuals - or groups that might include neurodivergent individuals. Review the statements in the left-hand column of Table 5.1 and tick the second column if you have ever performed that particular action.

Table 5.1 How neuroinclusive are your conversations?

Do you ever…	*Tick here if you do this*	*How this may be received*	*Instead try*
…make statements that imply something but don't explicitly say it, e.g. "I see you *finally* decided to join us" (which might imply that you are late) or "Are you wearing *that* to the meeting?" (which might imply that you are wearing something inappropriate)?		Literally!	Being direct, e.g. "The meeting started five minutes ago"; "The attire is collared shirts for men"
…use non-literal phrases, e.g. "Letting the cat out of the bag", "Kick the bucket", "It's raining cats and dogs", "It'll cost an arm and a leg"?		Literally! Someone may wonder why the cat was in the bag, or what the benefit of kicking a bucket would be	Using literal language, e.g. "It's raining hard"
…say things that aren't factually accurate, e.g. "I'll be back in five minutes" (meaning you will likely be back relatively soon but not specifically five minutes)?		That you will be back in *exactly* five minutes	Being specific with your language without assuming that people will understand nuances
…use body language to express an opinion, e.g. making an "unimpressed" face as a response to a comment?		The meaning of the body language may be missed or misunderstood	Being explicit about what you think, using verbal language

(Continued)

Table 5.1 (Continued)

Do you ever...	*Tick here if you do this*	*How this may be received*	*Instead try*
...say something sarcastic, i.e. saying the opposite of what you mean but expecting people to pick up on your intention because of the context, your voice and your facial expression?		Sarcasm may be missed or misinterpreted as passive-aggressive	Avoiding sarcasm and saying what you mean
...provide instructions verbally without any written follow-up (perhaps expecting others to take notes)?		Some may find the amount of information overwhelming and lose track	Providing written information in advance and/or slowing down your delivery; checking for understanding, e.g. "How does that sound?"
...speak fast in order to get through all your points?		Some may not be able to process what you are saying and therefore not be able to respond	Leaving more time so that everyone can process and fully understand what you are saying
...use your own preferred medium (e.g. phone, text), rather than that of your audience?		The audience may not be fully able to absorb the information you are providing	Considering your audience when choosing a medium, incorporating different options such as email, picture boards/visual cues, face-to-face and phone calls

Having done that, move onto these self-reflection questions:
Based on your responses in Table 5.1, have you noticed anything about how you communicate that could be more neuroinclusive?

....

....

....

....

What might you do differently in future?

....

....

....

....

The *who, what, when, how* and *where* of communication

The purpose of all communication is to share information with an audience in order to achieve an outcome (e.g. connect emotionally, exchange knowledge or problem-solve). Effective communication allows that audience to receive and absorb your message in the way you intended. As such, it needs to be tailored to the audience (*who*), include appropriate content (*what*), undertaken at the right time (*when*) and place (*where*) and in a constructive way (*how*).

Who

Understand your audience and who they are. Everyone can benefit from being asked how best to communicate with them, but this is all the more important for neurodivergent employees, given the differences in communication approaches already discussed.

The following questions can aid a better understanding of how to best communicate with your team. Even when you think you know the answer, ask directly anyway, and do so (at least) annually to ensure that you aren't making assumptions. Important as this is for neurodivergent individuals, we recommend you have the same discussions with neurotypical team members, too.

1. Do you have a preferred way of communicating or receiving information, e.g. email, text, face-to-face, audio, visual aids? (We are not implying that you always communicate in that one way with that particular individual: that may not be practical; rather, you are aware of which modes they find better or worse.)

2. How do you prefer to give and receive feedback, e.g. written, verbal, 1:1 meetings, frequency?
3. What do you need in order to be at your best in meetings? For example, would you prefer to be called on specifically to share your perspective? Or put up your virtual hand (on an online call) when you are ready to contribute?

The above questions will help provide you with a baseline understanding of your team members' communication preferences. All the same, be mindful when preparing each communication and reflect on what you know about your audience: what do they already know? What do they need to know? How might they perceive the message in the present context? What are their motivations and concerns, and how might you incorporate that into the message?

Where

Consider the appropriate medium and approach. Maybe a phone call works better in discussing a detailed topic rather than exchanging multiple emails. However, there are additional considerations where neurodivergent individuals are concerned; for example:

Written communication. Reading can be difficult, in particular for individuals with dyslexia. In the interests of clarity and inclusivity:

- Use clear sans-serif fonts (such as Arial or Calibri) at a minimum size of 12pt
- Increase the space in your documents: set loose line spacing within paragraphs and spaces between
- Structure your text using headings and subheadings
- Use bullet points to present information more concisely and break up paragraphs of text
- Avoid using all caps (e.g. for emphasis); use bold (not italic) for emphasis
- Use coloured backgrounds (e.g. cream or pastel) and dark blue type
- Remember the 3Cs (clarity, courtesy and conciseness)
- Avoid acronyms (or at least spell them out)

- Avoid figurative, ambiguous phrases, e.g. "circle the wagons"
- Don't use emojis

Face-to-face. In-person communication can be uncomfortable because of social anxiety, or discomfort with eye contact, for example. Walking and talking or sitting side by side can help to alleviate this. To reduce uncertainty (e.g. for those with OCD), it can be helpful to provide advance warning of meetings, and as much information as possible about the location and setup.

Online. Calls using a camera can be visually overstimulating. If a video call is required, allow participants the option of switching their camera off, or dialling in with their phone instead. It might be helpful in advance to set up speed dials for the numbers used, to help those with dyscalculia.

Verbal. Consider the tone, pace and rhythm of your communication, thinking about all the points discussed above.

- *1:1 versus group.* Group discussions can be difficult for some. For example:
    - Multiple voices talking over each other can be overstimulating
    - Knowing whose turn it is to speak next
    - Processing multiple viewpoints in quick succession can be challenging
    - Speaking up in a group can be intimidating for those who fear rejection or who lack self-esteem
- Typically, the larger the group, the more difficult it can be. Ways to mitigate these challenges include:
    - Turn-taking: encouraging everyone to speak by taking turns when appropriate
    - The "one, some and all" method: first, take time to reflect individually, then discuss with one or two others, before bringing thoughts to the main group. This approach helps those who may be hesitant to speak up to develop their thoughts and confidence

  - "Vote first then discuss". This model encourages different people's perspectives to be taken into account before starting a discussion. As leader, it's recommended not to include yourself in the vote as that may sway others' behaviour
  - Use a visual wall planner when discussing events and schedules
  - Provide reminders for appointments, and provide calculators in meetings. This can be helpful for those with dyscalculia
- *Sounds*. People differ in how they are impacted by background noise and in their sensory experiences. Some (e.g. with ADHD) find it easier to concentrate when there is some background noise; others (e.g. with autism) find it easier to have a quieter space. Adapt the space to your audience

What works for one person may not work for another. Furthermore, an individual's needs change depending on their internal state at the time. The key is to ask what would work best for the individual given both the present situation for them and the nature of the exchange. Simply ask "What would work best for you right now?"

When

Most people, neurodivergent and neurotypical alike, know that they are more focused at certain times of the day or week. So ask what them what is their best time to meet with you.

Preparation time is important, particularly if information needs to be digested beforehand, e.g. ahead of a performance discussion. Even regular meetings can be made less anxious by pre-sharing an agenda (including estimated timings and speaker names).

If you don't feel you have the energy to talk to an employee (let's say it's the end of a long day), it can be helpful to share your perspective, e.g. "I have a had a full day and I want to debrief you on the project, but I don't have the capacity to go into detail today. What would you like me to do: call you for ten minutes now or wait until tomorrow afternoon when you're in the office and we can speak in more detail?" This builds trust and encourages others to be open in similar situations.

What

Your communication comprises both the words and structure you use and the approach you take. This section covers the use of plain English - creating clear structure and content. We will then look at improving communication by building trust and active listening.

The ultimate aim of all communication is a clear and concise message that is consistently reinforced. The ideal vehicle for this is what we will call "plain English", which can be defined as communication that offers a "cooperative, motivated person a good chance of understanding it at first reading, and in the same sense that the writer meant it to be understood".[7] Not only is it intended to be easy for your reader, it is usually easier to write as well. Basic principles of plain English include:

- **Structure**
  - Keep your sentences short with one main idea per sentence
  - Use bullet points where possible to list key points or actions
  - Plan what you are going to say (i.e. your message, the structure) before you start writing
  - Lay it out for ease of navigation, i.e. with headings and demarcated sections
- **Language**
  - Choose short words over long words. Use words that your readers are likely to understand
  - Choose simple words over complex words, e.g. "details" in favour of "particulars", "about" over "regarding" and "end" over "terminate"
  - Use positive language in favour of negative language where possible, i.e. what to do rather than what not to do
  - If numbers are involved, just include those that are necessary, out of consideration for those with dyscalculia

Now look at the email shown in Box 5.2. What would have made the language clearer?

Box 5.2: Example email: First draft

Subject: Comprehensive Instructions for Project Plan Development

Dear Team,

It is paramount that we embark on the formulation of the project plan quickly, ensuring all requirements are thoroughly incorporated into each stage of the process, as detailed below. First and foremost, the initial phase should consist of collectively identifying all pertinent objectives, deliverables and stakeholders, while also meticulously outlining the expected scope, timeline and resources. The necessity for full participation in these preliminary discussions cannot be overstated.

Subsequently, the detailed delineation of roles and responsibilities, coupled with an exhaustive risk assessment, must be executed with precision to prevent unforeseen contingencies from derailing our trajectory. Teams are required to synergise their efforts toward the integration of comprehensive milestones and key performance indicators within the operational framework.

In addition, it remains imperative that all communication protocols are established, ensuring that interdepartmental collaboration is seamless, yet adaptive to fluctuating circumstances, in order to minimise inefficiencies. Finally, upon synthesis and finalisation of the draft plan, all participants are obligated to engage in rigorous review and feedback sessions to secure alignment and mitigate potential discrepancies.

Your attention to these steps is greatly appreciated.

Kind regards,
Bob

Next, see an alternative version, written in plain English (Box 5.3). Note the priorities given to:

- Short words over longer words
- Elimination of unnecessary words
- Bullet points over long sentences
- Short sentences over long sentences

Be aware of your own filters, too, as they can affect your clarity. Examples of filters are the biases that we covered in Chapters 2 and 3, our beliefs, our values, our morals and our past experiences. For example, someone who has had a bad experience of a performance review

Box 5.3: Example email: Second draft

Subject: Next Steps: Project Plan Development

Dear Team,

It is key that we create the project plan quickly. I've listed the next steps below:

1. Agree the key project criteria:
   a. Objectives
   b. Deliverables
   c. Stakeholders
   d. Scope
   e. Timeline
   f. Resources
2. Agree roles and responsibilities
3. Complete the risk assessment
4. Cross-check that milestones and key performance indicators are in sync between teams
5. Review as a group to ensure alignment

Thanks for your help.

Kind regards,
Bob

may feel nervous when the next one comes along. However, someone who didn't have that bad experience may find performance reviews motivating and rewarding. If both people are told their performance review is going to be "interesting", one may take that to mean that the review is negative and the other that it is positive.

Such filters can be helpful when it comes to communicating with someone with the same kinds of filters. But neurodivergent individuals' filters often differ to those of neurotypicals, because they are likely to have different ways of thinking about, interacting with and processing the world around them. Here, filters can inhibit clear and effective communication.

Filters play a part in our conversations usually without us noticing. If a team member says "I feel overwhelmed", you might take that to mean that they have too much work on their plate, as that is your understanding of the word. However, they might actually mean that they can't cope with the noise distractions in the office. And so, if you, as a manager, are applying your own filter, you may miss the opportunity to understand their problem and the opportunity to help them solve it. Perhaps you end up trying to solve their "overwhelm" by giving them less work rather than by enquiring what is in fact driving the overwhelm.

To reduce the risk of filters getting in the way of clear communication, use these two steps:

1. Reflect their words. Repeat your team member's exact words to them in order to demonstrate understanding and build rapport. For example, if they say, "I am exhausted by the amount going on", reply with: "You're exhausted. What can I do to support you?"
2. Ask neutral questions. These tend to be free from judgement or bias, and often start with "what" or "how". Avoid questions that start with "why", because these often elicit a defensive response. For example, consider the difference between "Why is the report late?" and "What led to the report coming in late?"

Try to avoid the interpretations and assumptions that arise as a result of your filters. One example is the interpretation of "overwhelmed", given above. Rein in the assumption that your team member wants you to solve their challenge; if, in conversation, you are not sure what they want from you, then ask: "how can I help you move forwards with this?" Avoid giving advice when what they actually want is a sounding board: someone to talk the problem through with, or simply to share it with for awareness.

Active listening

In everyday conversations, people are routinely listening in order to formulate a response, rather than to fully understand the meaning of the words being said. For example, when your co-worker describes their holiday, your mind may well immediately go to how that might apply to or inform your next holiday. You might not be listening to the details, nor reading the expression on your co-worker's face, nor noticing their tone of voice. Your co-worker might even notice the glazed look in your eyes as you daydream about your own travels.

Active listening, in contrast, involves fully focusing on the content, processing it and responding to make sure that you fully understand the meaning. It requires the listener to "get inside" the speaker in order to "grasp, from his point of view, what are they communicating to us[?]".[8] The key techniques within active listening are listed in Table 5.2.

Developing trust

Let's take an example of a company going through a redundancy process. Redundancies in themselves are stressful situations, creating uncertainty around critical parts of employees' lives. However, in companies in which the employees don't trust the decision-making process and their leadership team, those uncertainties and concerns are likely to create high levels of stress and fear. Chronic stress is thought to reduce productivity, and negatively impact relationships at home and at work;[9] fear is negatively related to productivity and performance.[10] Feeling scared or under threat in that way impacts your ability to be productive, creative and focused on your work,[11] as well

Table 5.2 Key techniques of active listening

Technique	*What this looks like*
Full attention	Giving the speaker your full attention. That means not getting distracted by other things going on (e.g. close your computer or put your phone away), and demonstrating this via body language (e.g. nodding) and eye contact (depending on the audience).
Reflective responses	Reflecting back, i.e. using the exact words the speaker has uttered, e.g. "I'm hearing that you're worried about the presentation for the upcoming meeting".
Empathy	Demonstrating empathy by acknowledging and validating the speaker's emotions, e.g. "I can see why that this is worrying for you". Empathetic managers have been shown to be more able to effectively support their team members.
Clarifying questions	Asking open questions to clarify understanding and progress the conversation, e.g. "Can you tell me more about what is worrying you about it?"

Source: Szulc, J.M., "Leading with understanding: cultivating positive relationships between neurotypical leaders and neurodivergent employees". *Journal of Employee Relations*, 46(9) (May 2024), 97-114. https://doi.org/10.1108/ER-12-2023-0621

as having a broader impact on interpersonal relationships and team dynamics. As a result of this lack of trust, company, team and individual performance can suffer.

It's not just in redundancy situations where trust in managers and leaders is important. Studies have shown that trust helps to improve creativity,[12] increase knowledge sharing and collaboration[13] and improve work satisfaction.[14] Trust is therefore critical to the success of any organisation, and is a key component of your role as a manager and leader.

This is confirmed by our understanding of the human brain. When it perceives a threat, it releases chemicals which activate the sympathetic nervous system; this readies our bodies for "fight or flight or freeze", i.e. to fight the threat, run away from it or freeze in front of it.[15] Indicators of this are increased heart rate or tense muscles ("fight"), rapid breathing, increased adrenaline ("flight") or a feeling of paralysis and numbness ("freeze"); it's an uncomfortable place to be. Sustained fight-or-flight response has been linked to long-term health impacts, such as high blood pressure,[16] heart disease,[17] reduced resistance to disease[18] and anxiety.[19]

However, when we find ourselves in the company of someone we trust, a different chemical is released, which calms the sympathetic nervous system, returning us to a more relaxed state.[20] This in turn makes us more able to focus and more able to think logically. As a result, productivity and personal relationships - and therefore team morale - all have an opportunity to grow. We will explore how to develop trusting relationships further in Chapter 6.

Common managerial situations

So far, we have focused on general communication techniques. However, managers and leaders are faced with specific situations in which clear, consistent and concise communication are particularly important; these include team meetings, feedback and delegation conversations.

Neuroinclusive meetings

There are simple actions that can make meetings more effective and inclusive of everyone. Applied in combination, they will reduce anxiety, improve the quality of input, allow sufficient time for preparation and make the meeting more effective.

Before we get to the checklist of actions, let's begin with another self-reflection exercise about your own experience of meetings.

Think about a meeting you've attended which you judged to be badly run. What happened?

....

....

....

....

What do you think led to it being badly run?

....

....

....

....

Now consider a well-run meeting you've attended. What was different about that meeting?

....

....

....

....

What was the impact of the meeting being well run? On you? On the group?

....

....

....

....

A checklist for meetings

This checklist is a tool to facilitate meetings that cause minimal social anxiety, that are an encouraging space for participation, and which allow people to follow along. It can enable more effective meetings for all neurotypes.

1. Before the meeting:
    - Ensure that the meeting is scheduled with sufficient notice to prepare
    - Ensure that you have a diversity of voices and perspectives in the room
    - Announce what medium will be used (e.g. Zoom, Teams), and give instructions on how to access it ahead of time
    - Provide the agenda ahead of time, including any preparatory reading in a variety of accessible formats. Allow for pre-submitted questions (where feasible) so attendees don't have to think of answers spontaneously. For example, if you're a having a career discussion, share the competencies in advance

- ○ Make written documentation easy to read and understand (see the above points on communication formats)
- ○ Set expectations. For example, say, "We want to hear from everyone in the room in turn". This can encourage less confident and more diverse voices to speak up

2. During the meeting:
   - ○ On video calls, allow for closed captioning and permit attendees to switch their cameras off (if possible)
   - ○ For in-person meetings, ensure that the environment is sensory-friendly (e.g. reduced noise and lighting)
   - ○ As attendees enter the room (real or virtual), welcome them all by name. Do not expect a response in all cases
   - ○ Ask everyone to have their full name up on screen, or name badges if in-person
   - ○ Set the tone. For example, say, "All ideas are welcome". (See Chapter 6 on creating a safe space)
   - ○ Use plain English (see above)
   - ○ Encourage everyone in the meeting to share, if they are comfortable in doing so. For example:
     - Say, "Is there anyone we haven't heard from so far who would like to contribute?"
     - Encourage use of the "hand raise" function on video calls
     - Ask attendees to take turns speaking: you could go round the (real or virtual) room in turn and ask for input
     - Encourage attendees to engage using alternative media such as chat functionality, gestures or animations
   - ○ Allow for comfort breaks in meetings longer than one hour. This helps attendees to stay focused. Allow them to move around the (real or virtual) room, too, for the same reason
3. After the meeting:
   - ○ Provide a written summary. This will usually include actionable items with their due dates and accountable owners. This can be AI-generated

Neuroinclusive delegation

We've just seen how neuroinclusivity improves the quality of meetings for *everyone*. The same applies to neuroinclusive delegation. Some managers have difficulties in delegating generally: they struggle with the idea that the work might not be done the way they would do it or to their standards. Sometimes, when they think their concerns were justified, they will reassume control of the task, sacrificing learning opportunities in the process. There are simple techniques and approaches to tackle both of these common challenges.

Firstly, know your team. Be aware of what tasks they are currently working on, which they would like to be working on, their strengths and challenges, and their ambitions. Ideally, tasks delegated should fit with their broader ambitions, strengths and learning goals. If you don't know what those are for a member of your team, ask them. See more on identifying strengths in Chapter 7.

Secondly, start small. Build trust in your team by delegating straightforward tasks, and seeing how they go, giving feedback along the way as needed (even if you have a recurring meeting in the calendar for that purpose). This helps your team to build confidence in their own abilities.

Thirdly, be clear about your expectations. Pre-empt the question "What do you need, by when and why?" Be specific (and reiterate what has been agreed in a meeting in writing). Managers often miss out the "why" part and fail to provide context, which can be particularly important for neurodivergent individuals. Agree progress check-ins, which should give you the information you need, but resist the urge to micromanage along the way. Focus on the end goal, and don't (however tempting) mandate the employee to perform the task exactly how you would do it; their way, assuming it meets the end goal, might be just as good if not better than yours. Taking different routes is a recognised trait of many neurodivergent individuals.

Lastly, give and receive feedback through the process, so that both you and your direct report can understand how to improve the way you work with each other.

Neuroinclusive feedback

The very idea of "being given feedback" can fill some people with dread, the assumption being that it will be critical. It can trigger the "fight or flight or freeze" mechanism, where the heart beats faster, or the brain goes into a panic, obsessing about the impending meeting. Neurodivergent individuals, as we learned, are more likely to have experienced trauma than neurotypical individuals, and more likely therefore to perceive feedback as a threat and be triggered in this way.[21] The effective provision of feedback is always going to be a key part of a manager's role, but when neurodivergent individuals are involved expertise in this area is all the more important.[22]

However, it's not only neurodivergent team members that can respond emotionally to feedback. Neurotypical and neurodivergent managers, too, can respond emotionally to a situation. Preparation is key. Try to centre yourself before the conversation. Do some mindful breathing, for example, or take time to focus on what emotions you are feeling and what they are connected to (see Chapter 12 on resources). Ask yourself what is driving your expectations around the other person's performance.

Consider having a recurring meeting in the diary which, it is understood, will comprise both positive and constructive feedback. This avoids the kinds of spontaneous announcement that can trigger the fight-flight response.

Think about the physical environment, the aim being to create a psychological safe space (see Chapter 6). Find somewhere where you cannot be overheard or interrupted, to ensure confidentiality for all concerned. Ask not to be disturbed. In online meetings, earphones can be used if the conversation might be overheard, and notifications switched off.

In preparing the actual content of the feedback, a useful model is the Action-Impact-Do Differently (AID) model, as described in the following list, which can be used for both constructive and positive feedback, in tandem with the strengths-based approach described in Chapter 7. The model helps make the feedback actionable and clear.

- **Action**. Describe precisely what you are providing feedback on: focus on the employee's actual *action*. Ensure the feedback is as objective and as factual as possible. In articulating the action taken, phrase it in a non-confrontational and neutral way. For example, "The report was sent in after the 5 pm deadline" is more factual and precise than "You sent the report in late"

- **Impact**. Describe the impact of the action on the team or the project. Again, stick to facts rather than thoughts or feelings you might have about the employee. This maintains focus on what is important about the action. For example, "This meant that the team could not progress with the next task, putting us behind our deadlines overall" or "Your approach allowed me to think differently about our products and come up with something which we hadn't previously thought of". Check for understanding: ask them for their interpretation, pausing occasionally to summarise to ensure you are all in sync
- **Do/do differently**. What would you like the person to continue doing (positive feedback) and/or change (constructive feedback)? For example, "In the future, I would like to know at least a day in advance if you are unable to meet the deadline". In this stage, particularly for constructive feedback, ask questions to ensure that the desired action is possible, e.g. "How does that sound to you?", "What else might you want to do to ensure that the deadline is met next time?" This is particularly important for ND individuals, who are more likely to have a different perspective, meaning a greater potential for misunderstanding or difference in viewpoints. Perhaps lay out a specific action plan based on the feedback, so the employee has a structure to facilitate their change of behaviour

ND individuals may believe that they have to prove themselves,[23] and can struggle with self-esteem,[24] so find a way to appreciate them and acknowledge how they are contributing. Ask what form they would like that acknowledgement to take: some may prefer positive feedback delivered privately face-to-face rather than in a public forum or from a senior manager. They may also not interpret the feedback the way it was intended, i.e. positively.

Finally, to reiterate: feedback conversations should not be anxiety-inducing infrequent events, saved for performance reviews. Feedback should be ongoing and bidirectional.

Box 5.4: Case studies in communication

These examples share some managers' do's and don'ts when it comes to communicating with their team members. They highlight the importance of understanding people's expectations about meetings, what useful feedback looks like and the difference you can make in easing employees' stress by simply checking in.

> ...lack of consistency in communicating important changes/being given information/agenda before a meeting where I know I'll have to contribute, not knowing in what capacity (not setting clear expectations nor being given agenda points). This often leads to increased stress and worrying. The reality is that I receive very positive constructive feedback, as long as this leads to an open conversation, feedback is evidence-based and it focuses on how I can improve my performance in certain areas, as opposed to being told what hasn't worked in a more generic way. Therefore, feedback is always appreciated although at times managers haven't given me any as they assumed I'd react negatively because they read somewhere that people with ADHD have difficulty in accepting feedback altogether.
>
> 39-year-old male EDI in higher education with ADHD

> Listening has been the most positive behaviour, along with looking at ways to provide clarity, prioritising workstreams and checking in with me as to which meetings/group sessions I feel I can manage and which I might need to build up to or avoid for the time being.
>
> 58-year-old male in the healthcare sector who is AuDHD

Neuroinclusive emails and chat messages

Based on what you have learned in this chapter about neuroinclusive written communication, consider how the email in Box 5.4 might be changed to make it more neuroinclusive. Having read the email, reflect on the following:

Box 5.5: Example email #2

Hi Team,

I hope everyone had a great weekend. As you all know, we have a PROJECT DEADLINE COMING UP, and we need to make sure everyone is pulling their weight. We need to be on the ball this week to get everything completed on time!

Please make sure you all attend the team meeting tomorrow at 9 am where we are aiming to get on the same page about progress. No excuses. It's mandatory for everyone.

Also, I noticed that some of you have been slacking off and taking longer breaks than usual. Let's try to stick to the 15-minute break policy, as we need to stay focused.

If you have any questions or concerns, let me know.

Thanks,
Bob

What are the actual facts in the email? What meaning has the author created from those facts?

....

....

....

....

Which words might mean different things to different people?

....

....

....

....

What impact might that have on the email's recipients?

....

....

....

....

How could you change the email to make it more neuroinclusive?

....

....

....

....

Other common communication pitfalls

Finally, we will round this chapter off with some common communication pitfalls that we haven't covered yet, and which managers and leaders can fall into on account of their biases and assumptions.

1. **Excessive support**. You can easily fall into a pattern of repeatedly offering a neurodivergent individual help and support, even when they haven't asked for it. This is particularly true when you're involved in providing daily oversight. A manager might believe they are being caring and supportive but the employee may experience this very differently and believe they are being seen as helpless or being micromanaged. Advise your team member what level of support can be offered and ask how they would like to check in with you.
2. **Commenting on neurodivergent traits**. Examples are: "I'd never have known you were autistic" or "Don't trip on the step, those of you who are clumsy". Managers may well think they are being helpful, or light-hearted, but implying that certain individuals conform to stereotyped neurodivergent behaviour can prompt them to hide their differences, or else make them angry about the assumption.

3. **Assuming an employee knows what they need.** If an employee is new to their working environment, for example, they might not yet be in a position to say. In that situation, a statement like "If you need anything, let me know" isn't helpful. Instead, you could provide both a written and verbal list of support options, and how to access them.

Key points from Chapter 5

1. Adopt the 3Cs - clarity, courtesy and consistency - as the simple rule of communication.
2. Always consider:
   a. *Who*. Your audience, and their preferences and current levels of understanding
   b. *What*. The right message
   c. *Where*. The appropriate environment and medium
   d. *When*. Timing the message to ensure that it lands as intended
   e. *How*. Active listening and plain English
3. Use the checklist for making meetings more inclusive for everyone, which includes sending an agenda and relevant materials ahead of time.
4. Be thorough in preparing for delegation and feedback conversations so that your communication is encouraging, clear and concise - and has the impact that you want.
5. Being curious about others' perspectives, and being willing to learn, is a great place to start for all communication.

Notes

1 Mastroianni, K., and Storberg-Walker, J. "Do work relationships matter? Characteristics of workplace interactions that enhance or detract from employee perceptions of well-being and health behaviours". *Health Psychology and Behavioral Medicine*, 2(1) (1 January 2014), 798–819. https://doi.org/10.1080/21642850.2014.933343
2 Clampitt, P.G., and Downs, C.W., "Employee perceptions of the relationship between communication and productivity: A field study". *Journal of Business Communication*, 30(1) (1993), 5–28. https://doi.org/10.1177/002194369303000101.
3 Goulet, J.D., "Stop asking neurodivergent people to change the way they communicate". *Harvard Business Review*, 5 October 2022. Available at: https://

hbr.org/2022/10/stop-asking-neurodivergent-people-to-change-the-way-they-communicate

4 Maxwell, K., *The 17 Indisputable Laws of Teamwork: Embrace Them and Empower Your Team*. London: HarperCollins, 2013.

5 Goulet, J.D., "Stop asking neurodivergent people to change the way they communicate". *Harvard Business Review*, 5 October 2022. Available at: https://hbr.org/2022/10/stop-asking-neurodivergent-people-to-change-the-way-they-communicate

6 Lange, S. "Different not less: Neurodiversity as a lens for understanding our students better". *Scope Contemporary Research Topics (Learning and Teaching)*, November 2022. https://doi.org/10.34074/scop.4011005. Ford, I., *A Field Guide for Earthlings*. Albuquerque, NM: Ian Ford Software Corporation, 2010. Garrels, V., "Getting good at small talk: Student-directed learning of social conversation skills". *European Journal of Special Needs Education*, 34(3) (2018), 393–402. https://doi.org/10.1080/08856257.2018.1458472

7 Cutts, M., *Oxford Guide to Plain English*. Oxford: Oxford University Press, 2010.

8 Rogers, C., and Farson, R.E., "Active listening". In Newman, R.G., Danzinger, M.A., and Cohen, M. (eds), *Communicating in Business Today*. Lexington, MA: D.C. Heath & Company, 1987.

9 Timotius, E., and Octavius, G., "Stress at the workplace and its impacts on productivity: A systematic review from industrial engineering, management, and medical perspective". *Journal of Industrial Engineering and Management Systems*, 21(2) (June 2022), 192–205. https://doi.org/10.7232/iems.2022.21.2.192

10 Pustovit, S., Miao, C., and Qian, S., "Fear and work performance: A meta-analysis and future research directions". *Human Resource Management Review*, 31(3) (2024). https://doi.org/10.1016/j.hrmr.2024.101018

11 Pustovit, S., Miao, C., and Qian, S., "Fear and work performance: A meta-analysis and future research directions". *Human Resource Management Review*, 31(3) (2024). https://doi.org/10.1016/j.hrmr.2024.101018

12 Klimoski, R.J., and Karol, B.L., "The impact of trust on creative problem solving groups". *Journal of Applied Psychology*, 61(5) (1976), 630–33. https://doi.org/10.1037/0021-9010.61.5.630

13 Chen, Y.-H., Lin, T.-P. and Yen, D.C., "How to facilitate inter-organisational knowledge sharing: The impact of trust". *Journal of Information and Management*, 51(5) (2014), 568–78. https://doi.org/10.1016/j.im.2014.03.007

14 Lee, G., and Teo, A., "Organizational restructuring: Impact on trust and work satisfaction". *Asia Pacific Journal of Management*, 22 (2005), 23–39. https://doi.org/10.1007/s10490-005-6416-6

15 Roelofs, K., "Freeze for action: Neurobiological mechanisms in animal and human freezing". *Philosophical Transactions of the Royal Society B: Biological Sciences*, 372 (27 February 2017). https://doi.org/10.1098/rstb.2016.0206

16 Lohmeier, T.E., "The sympathetic nervous system and long-term blood pressure regulation". *American Journal of Hypertension*, 14(S3) (June 2001):147S–154S. https://doi.org/10.1016/S0895-7061(01)02082-9

17 Florea, V.G., and Cohn, J.N., "The autonomic nervous system and heart failure". *Journal of Circulation Research*, 114(11) (23 May 2014). https://doi.org/10.1161/CIRCRESAHA.114.30258

18 Kenney, M.J., and Ganta, C.K., "Autonomic nervous system and immune system interactions". *Comprehensive Physiology*, 4(3) (July 2014), 1177-200. https://doi.org/10.1002/cphy.c130051

19 Hoehn-Saric, R., and McLeod, D.R., "The peripheral sympathetic nervous system: Its role in normal and pathologic anxiety". *Psychiatric Clinics of North America*, 11(2) (1988), 375-86.

20 Mandolfo, M., "You trust me and I feel it. Influence of foreign live biofeedback on interpersonal trust-related behaviour". *Milan Polytechnic Library Archive*, 2015/16. Zak, P.J., "The neurobiology of trust". *Scientific American*, 298(6) (1 June 2008), 88-92. https://doi.org/10.1038/scientificamerican0608-88.

21 Hoover, D.W., "The effects of psychological trauma on children with autism spectrum disorders: A research review". *Review Journal of Autism and Development Disorders*, 2 (9 July 2015), 287-99. https://doi.org/10.1007/s40489-015-0052-y. Lobregt-van Buuren, E., Hoekert, M. and Sizoo, B., *Autism Spectrum Disorders*. Brisbane, QL: Exon Publications, 2021; Szymanski, K., Sapanski, L., and Conway, F., "Trauma and ADHD - association or diagnostic confusion? A clinical perspective". *Journal of Infant, Child, and Adolescent Psychotherapy*, 10(1) (2011), 51-59. https://doi.org/10.1080/15289168.2011.575704. Alexander-Passe, N., "Investigating Post Traumatic Stress Disorder (PTSD) triggered by the experience of dyslexia in mainstream school education". *Journal of Psychology and Psychotherapy*, 5(6) (January 2015), https://doi.org/10.4172/2161-0487.1000215. Wadsworth, L.P., Van Kirk, N., August, M., et al., "Understanding the overlap between OCD and trauma: Development of the OCD trauma timeline interview (OTTI) for clinical settings". *Current Psychology*, 42 (2023), 6937-47. https://doi.org/10.1007/s12144-021-02118-3. Van der Kolk, B.A., "Clinical implications of neuroscience research in PTSD". *Psychobiology of Posttraumatic Stress Disorder: A Decade of Progress*, 1,071(1): 277-93. https://doi.org/10.1196/annals.1364.022. Van der Kolk, B.A., "The neurobiology of childhood trauma and abuse". *Journal of Child and Adolescent Psychiatric Clinics*, 12(2) (April 2003), 293-317. https://doi.org/10.1016/s1056-4993(03)00003-8. Van der Kolk, B.A., "Posttraumatic therapy in the age of neuroscience". *Psychoanalytic Dialogues*, 12(3) (2002), 381-92. https://doi.org/10.1080/10481881209348674

22 Szulc, J.M., "Leading with understanding: Cultivating positive relationships between neurotypical leaders and neurodivergent employees". *Journal of Employee Relations* 46(9) (May 2024): 97-114. https://doi.org/10.1108/ER-12-2023-0621

23 Szulc, J.M., "Leading with understanding: Cultivating positive relationships between neurotypical leaders and neurodivergent employees". *Journal of Employee Relations* 46(9) (May 2024): 97-114. https://doi.org/10.1108/ER-12-2023-0621

24 Buckley, E., Sideropoulos, V., Pellicano, E. and Remington, A., "Higher levels of neurodivergent traits associated with lower levels of self-efficacy and wellbeing for performing arts students". *Neurodiversity*, 2 (13 April 2024). https://doi.org/10.1177/27546330241245354

6 Creating an environment in which everyone can thrive

As a team leader and manager, what you do and say impacts the tone and culture of your team. In the last chapter we looked at your approach to conversations; so now let's look at how you can create the best possible *culture* in your team – one that is truly inclusive of all individuals. In this chapter, we will see how your behaviours define and shape team culture, we will examine what an inclusive leadership style could look like for you and we will identify which behaviours can create greater psychological safety.

How would you define your current team culture?

What words describe your team's culture right now? For example, is the team "lively", "fun", "siloed", "hard-working", "goal-oriented", "ethical"?

....

....

....

....

How have your language, behaviours and inactions as a leader and manager contributed to the team's current culture?

....

....

....

....

DOI: 10.4324/9781003645535-6

The behaviours of an inclusive leader

Perhaps the above self-reflection helped you appreciate that your behaviours, language and actions play a part in setting the culture and tone for your team. So, if you are to create a culture that welcomes and includes difference, what behaviours do you need to exhibit?

Deloitte have addressed this question through their "6 Cs of Inclusive Leadership" model:[1]

Commitment. Inclusive leaders demonstrate they are committed to the principles of diversity and inclusion. They believe that the value created through inclusion is worth it, and inclusion is one of their personal values. By demonstrating this commitment to themselves and others, they can inspire others' inclusive behaviours. These behaviours include:

- Understanding and communicating each team member's uniqueness. This encourages neurodivergent team members to feel valued
- Making inclusion a business priority
- Treating all team members with fairness and respect
- Adjusting your leadership style and behaviours to reflect your team members' requirements, e.g. communication style, working approach

Cultural intelligence. Having cultural intelligence means being able to identify your own cultural background and that of others and seeking to understand how cultural stereotypes can influence your own and others' expectations of others. This is strongly linked to curiosity (another of the 6 Cs; see below): the more culturally intelligent you are the more curious you are about other people's perspectives or frames of reference. Behaviours that demonstrate and develop cultural intelligence include:

- Being curious about other cultures and perspectives
- Looking for opportunities to spend time in culturally different environments
- Being considerate of the communication differences between

cultures, e.g. differences in personal space, pace, volume and style of communication

Courage. Inclusive leaders are not afraid to challenge the way that things are done, thereby addressing such issues as group-think,[2] unhelpful beliefs or unconstructive behaviours. They are willing to have uncomfortable or difficult conversations. Behaviours that demonstrate courage include:

- Seeking the help of others if you are not able to complete a task on your own
- Openly admitting when you have made a mistake
- Challenging organisational behaviours that are non-inclusive

Cognisance of bias. Inclusive leaders identify and challenge their own biases, and put in place practices to prevent biases from impacting team dynamics or decision-making. (More information on neurodiversity-related biases is in Chapter 2.) These behaviours include:

- Asking for feedback around your own personal biases
- Focusing on fairness and equity when making decisions around pay or promotions
- Identifying and challenging organisational processes that are inconsistent with merit, or which are inherently biased

Curiosity. Having an open mind and a desire to understand another person's perspective. To feed this curiosity, inclusive leaders respectfully ask questions and listen attentively to the answer, refraining from making quick judgements. Behaviours that demonstrate and develop curiosity include:

- Actively seeking the perspectives of others who have different perspectives
- Listening when another person is offering their opinion
- Looking for opportunities to connect with others who are different, i.e. in age, education or social status

Collaboration. Inclusive leaders are strong team players and believe that team collaboration is a key to performance and success. They create a psychologically safe space (see below) where people can bring all their ideas and perspectives without fear of judgement. Behaviours that demonstrate and develop collaboration include:

- Empowering team members to make decisions about issues that impact their work
- Recruiting teams made up of different perspectives, experiences and backgrounds
- Inviting diverse perspectives in discussions
- Creating a psychologically safe environment in which everyone is willing and able to speak up

The above leadership qualities need to be applied in different contexts, i.e. with both face-to-face and remote-working employees. For example, in hybrid meetings ensure that all team members have an equal opportunity to contribute.
Of the above behaviours, which ones do you already do consistently?

....

....

....

....

Which behaviours do you need to develop further in order to be more inclusive? What steps might you take to do that?

....

....

....

....

The last behaviour, "Collaboration", is particularly important when working in groups and teams. So how can you create the kind of environment described? We will now look at team activities to create an inclusive environment before investigating the idea of "psychological safety".

Team activities to help develop an inclusive culture

The below activities can support development of an inclusive culture:

- Ask team members in turn to share stories about when they've seen others in the team at their best, and what impact that had on them. For example, "in the client meeting the other day, your energy and enthusiasm about our product came across to the client". This contributes to a positive and supportive team environment. Give team members some time to prepare, so they can bring meaningful and considered points, and aren't pressured to answer on the spot
- Share origin stories. Origin stories summarise the first step in a relationship or a career step. For example, they might answer the question "When did we first meet?" or "What was it that brought you to this city?" or "What led you to choose this career?" This develops bonds by creating a shared understanding and points of connection. Consider even a recurring slot for these stories in team meetings
- Create rituals. For example, spend five minutes at the start working together on a lateral-thinking puzzle, or everyone shares something they have experienced. This can be a moment of lightness in the day and create a shared team identity
- Use the "one, some, and all" method of problem-solving: first reflect individually, then discuss in small groups of two or three, before bringing your thoughts to the main group. This boosts the confidence of those who may be hesitant to speak up in larger groups

Your brilliant idea

Imagine that your team has been working on a commercial problem for a few months. None of the many potential solutions have worked out so far. One morning, on your way to work, you have a brilliant, creative idea. You think it *might* work, but you're not sure. It's quite different to anything else the group has tried so far, and you're wondering how your team will receive it. As you approach the office, you're getting increasingly nervous about sharing your idea. They'll probably see it as unworkable. They've never considered this kind of idea before so what would make today different? Going in to the team meeting, you decide there's too much risk of embarrassing yourself, so you decide to not share it. They might just laugh at you and your weird idea.

Reflect on the following questions:

Have you ever been in a situation where you held back because you didn't think it was safe to speak up? When? What was the situation?

Have you ever had a dramatically different idea and shared it with your colleagues? When? What was the situation?

What led you to speak up in one situation but not in the other? For example, did the culture differ between the two situations? How much trust did you have in your colleagues in either space? What did you think about your team members? How accepted did you feel as an individual in each scenario?

What is psychological safety?

As the above reflection may have revealed, there are environments and situations in which people are more likely to share their ideas, and those in which they are less likely to do so. In the latter type, there is often a perceived threat to their identity, reputation, beliefs and/or comfort. In other words, there was a low level of "psychological safety".

Psychological safety is a shared belief, among a group of people, that a team is safe for interpersonal risk-taking.[3] When team members have psychological safety, they feel that the team will not embarrass, reject or criticise them for speaking up, making mistakes, raising concerns, seeking help or asking questions. Level of psychological safety is strongly correlated with the behaviour of the team's leader,[4] whose actions, inactions and words set an expectation of safety or lack thereof.

In teams with low levels of psychological safety, members will hold back from contributing because they believe they will be criticised. They may fear reputational damage or being judged as incompetent. They will not share their mistakes, believing that team members will punish them. They will not share alternative perspectives in case others feel embarrassment or discomfort. Ultimately, the *perceived* potential cost of sharing their thoughts is greater than the personal or organisational benefits.

In contrast, *psychologically safe* cultures are characterised by a willingness to have challenging discussions. Some indications that you are in a psychologically safe culture are that:

- Team members are able to give and receive feedback, including to/from their team leader
- They will ask each other and their leader for help when required
- Everyone will admit errors and seek to learn from them
- They will disagree with each other and their leader
- Members can be authentically themselves, sharing their strengths and challenges openly
- They can raise concerns about behaviours they witness in the team and elsewhere in the organisation

The benefits of such an environment are significant, not just for neurodivergent individuals, but for all employees. For example:

- **Improved innovation.**[5] When people feel it is safe to share their ideas, greater creative problem-solving and innovation ensue
- **Increased engagement.**[6] When people contribute, they are more personally committed
- **Stronger team engagement.**[7] Open, honest dialogue creates a stronger and more resilient team
- **Higher-performing teams.**[8] Teams with higher psychological safety demonstrate more openness and learning, and perform better as a result
- **Improved decision-making**[9]
- **A culture of continuous learning**[10]
- **Lower turnover.**[11] People enjoy learning and being accepted in psychologically safe cultures

Amy Edmondson, who originally coined the term "psychological safety", defined seven statements to determine a team's degree of psychological safety.[12] So look at the following seven statements, and consider in what situations each statement is true, partially true, sometimes true or not true, for your team. Having done that based on your own experience, it can be illuminating to ask some team members about their ratings and compare them with yours.

1. **If you make a mistake on this team, it is not held against you.**
   This statement is: True Partially true Sometimes true Not true
2. **Members of this team are able to bring up problems and tough issues.**
   This statement is: True Partially true Sometimes true Not true
3. **People on this team sometimes accept others for being different.**
   This statement is: True Partially true Sometimes true Not true
4. **It is safe to take a risk on this team.**
   This statement is: True Partially true Sometimes true Not true

5. **It isn't difficult to ask other members of this team for help.**
   This statement is: True Partially true Sometimes true Not true
6. **No one on this team would deliberately act in a way that undermines my efforts.**
   This statement is: True Partially true Sometimes true Not true
7. **In this team my unique skills and talents are valued and utilised.**
   This statement is: True Partially true Sometimes true Not true

Levels of psychological safety

According to Dr Timothy Clark,[13] there are four stages of psychological safety, which are based on how much permission is given to team members and how much respect they experience. As levels of permission and respect increase in tandem, so increases the level of psychological safety.

In reality, teams don't progress through each stage in a linear fashion. In fact, people will experience different levels of psychological safety at different times and in different situations. Team members may differ in how safe they feel. While all members are part of creating a psychologically safe environment, the leader sets the conditions for it to exist. Leadership practices to develop psychological safety will be presented later in this chapter.

Clark's four levels of psychological safety are listed below. Consider the teams with which you work and where they belong on the scale.

Level 1: Inclusion safety

There is respect for individuals as human beings, and people feel safe to belong to a team. They are accepted for who they are, and for their unique attributes and defining characteristics.

At this level of safety, people can bring their whole selves to work, e.g. they do not need to mask their neurodivergent traits (see Chapter 2).

Level 2: Learner safety

People can ask questions with the aim of learning and growing as they feel safe enough to share their level of competence. They are able to fail, and are encouraged to learn from it. Feedback is shared freely, in the spirit of learning from experience.

At this level of safety, neurodivergent individuals might share their personal strengths and challenges with the team, confident that their co-workers will accept them and support their learning and growth.

Level 3: Contributor safety

Members feel safe to contribute ideas that fall outside their job description, without fear of stepping on someone else's toes or having to compete with others.

At this level of safety, people are allowed to contribute in order to make a meaningful difference to the team. Neurodivergent individuals can share their ideas freely and increase the team's overall level of creative thinking.

Level 4: Challenger safety

Members feel safe to speak up, challenging norms and authority. In fact, this is encouraged, and there are no negative consequences of doing so. Leaders at this level encourage dissenting views and debate.

Which level is familiar to you and the teams you are in?

....

....

....

....

What do you think contributes to this? What could you do differently to create a more inclusive culture in these teams?

....

....

....

....

A high level of psychological safety sounds like a great thing. However, there are contexts where that is not the case.[14]

Imagine you are an air traffic controller. Your role in the team is to handle different planes coming in to land. This is probably not a context in which to take risks and relinquish your fear of failure, e.g. experimenting with how quickly planes might be able to land after one another. In roles like these, in which the consequences of making mistakes are serious, and in which there are clearly defined standards of behaviour, psychological safety is not necessarily correlated to performance.[15] As leader, consider the level of risk you are working with and the tolerance for risk-taking mistakes.

What psychological safety isn't

The three main misconceptions about psychological safety are:

1. It's about being "nice" to each other. Not necessarily. Psychological safety enables people to be candid and to challenge one another. If being nice is your priority, you will likely avoid the uncomfortable conversations.
2. Everyone must feel comfortable. If you are being challenged, taking risks and making mistakes, then you may well *not* feel comfortable. In fact, being outside of your comfort zone, you may feel vulnerable and anxious - but, crucially, you know that you have permission to be candid.
3. It's to the detriment of performance. This comes from the false perception that psychological safety means people aren't being held accountable for their performance. However, in a psychologically safe environment, people are receiving feedback, they take responsibility for their actions and are required to improve performance towards team goals. This improves the team's overall performance as well as that of the individual.

What makes psychological safety critical for neurodivergent individuals?

As we have seen, psychological safety can improve innovation, engagement, risk management and team cohesion. Neurodivergent team members in particular benefit from this because:

1. They are more likely to have different ways of thinking and therefore are more likely to be bringing the creative ideas forwards. If they do not feel safe, they may withhold their unique perspectives.
2. They can suffer from rejection sensitivity dysphoria (RSD) (see Chapter 2), making them more sensitive to perceived rejection. Increasing their level of safety can help to reduce the frequency and impact of RSD symptoms.
3. Studies have shown that neurodivergent individuals are more likely to have a more sensitive nervous system.[16] This means that they are more likely to experience threat responses if they don't feel they are in a safe and trusting environment. A psychologically safe environment therefore allows them to maintain a calm system.
4. Inclusive leadership, of which psychological safety is a key component, has been linked to reduced avoidance of the workplace by neurodivergent individuals. Perhaps this is because being in a psychologically *unsafe* environment expends additional energy ruminating about work or other worries driven by uncertainty and fear. Psychological safety might allow them the energy they need to go into the workplace.[17]

Psychological safety: practices

What are the behaviours that you, as a leader and manager, can exhibit to support a culture of psychological safety? Below we suggest some options. We suggest picking a few and role-modelling them consistently in order to embed them into the team's culture and ways of working, then reflecting with the team on how they are experiencing the cultural shift that may have been created.

1. According to Edmondson and Hugander,[18] creating a focus on outcomes helps people understand why it is important to speak up. In explaining the desired outcomes, be explicit about how their diverse perspectives and creative ideas will help contribute to that outcome, and the impact that they can have personally.
2. Actively invite input. For example, say: "What do you think about this?", "Where do you stand on this idea?"
3. Be authentic. For example:

    a. Remain accessible and approachable.
    b. Seek out feedback about the meaning people make of your own behaviours.
    c. Be honest in relaying information with staff, no matter how difficult you find it.
    d. Admit when you don't know something; make it OK to not know everything. Ask questions to gain others' knowledge; reiterate that it is the collective power of the group that makes the team successful.
    e. Admit your own mistakes and failings. Your example helps others feel OK about admitting theirs.

4. When people admit their mistakes, thank them. Appreciate the honesty and courage that it takes to own up to errors. For example, say: "Thank you for raising this; I appreciate the effort it may have taken to share this issue with me. What have you learnt from it?" Encourage people to raise mistakes and "bad news"; for example, start meetings with: "What is the news you think that I don't want to hear?"
5. When people put forwards ideas, acknowledge their ideas and explore them through curiosity, being aware of your assumptions. For example, "Tell me more about that", "That sounds interesting: what benefits do you think it might bring?"
6. Actively listen to your team members (see Chapter 5 on active listening). Some open questions to try are: "What do you want to accomplish?" or "What is concerning you at the moment?"

7. Frame meetings as opportunities to share information and different perspectives.
8. Encourage open discussion of strengths and weaknesses of different team members; make it OK to not be perfect.

What would you do?

Reflect on what you would do and say in the situations below.

Scenario 1: Your team manages payments to clients. One of your team members tells you that they have accidentally paid half a million pounds to the wrong client account. They are panicked, concerned that the money won't be returned and that the customer whose money it is will complain.

What do you do?

....

....

....

....

What do you say first?

....

....

....

....

Scenario 2: Your team manages payments to clients. One of your team members mentions in passing that they have accidentally paid half a million pounds to the wrong client account. They are relaxed about it; they are working on it but "it's not top of my list".

What do you do?

....

....

....

....

What do you say first?

....

....

....

....

Scenario 3: Your team manages payments to clients. One of your team members mentions that last week they had accidentally paid half a million pounds to the wrong client account because they weren't fully focused on the task. But they have since retrieved it and paid it to the right client.

What do you do?

....

....

....

....

What do you say first?

....

....

....

....

Scenario 4: Your team manages payments to clients. One of your team members tells you that they have accidentally paid half a million pounds to the wrong client account. This is the third time they have done this in the last month.

What do you do?

....

....

....

....

What do you say first?

....

....

....

....

What difference do you notice between your responses to the four scenarios?

....

....

....

....

How might you change your responses if you wanted to build psychological safety within your team?

....

....

....

....

As the above exercise may have highlighted, it is tricky to find the language and action that balances a need for performance and high standards alongside a need to enable people to make and raise mistakes. It takes time and thought. In situations like this, thorough preparation is beneficial.

Case study: Trust and Psychological Safety

The following statement speaks to the role of trust in enabling team members to be open about their challenges.

> My manager has said it's been helpful for me to be open with him, about the why and the how of work. He's encouraged me to be as open as I can about some complex health issues. This has created a relationship of trust, which helps with honesty, especially on days where I am struggling to be effective at work and may need downtime.
>
> 40-year-old female with ADHD,
> office manager for a small business

We now look in further detail at how to develop trusting relationships within your team.

Trust

As the above example demonstrates, trust is a cornerstone of psychological safety. Without trust, team members will not feel safe to express themselves openly. While psychological safety extends beyond trust, trust complements psychological safety by creating the interpersonal foundation on which it is built.

A Harvard research study by Frances Frei and Anne Morriss[19] identified three drivers of trust:

1. **Authenticity**. People feel they are interacting with the "real" you and that you have integrity.
2. **Logic**. People believe in the logic and good reasoning behind your decision-making.
3. **Empathy**. You treat people with care and respect.

Authenticity

Authenticity is about being true to who you are and being genuine in your interactions. To demonstrate authenticity:

- Be open and honest in sharing your challenges and what you believe (but only where appropriate as a leader, e.g. don't share your concern about upcoming redundancies as that may create more fear!). Being open in this way can help you to identify similar points of connection with others
- Admit mistakes. This demonstrates integrity and reinforces psychological safety (see above)
- Show vulnerability. This builds connections with others[20] and demonstrates that you trust them, which can result in reciprocated trust[21]
- Be present. Engage fully in conversations and demonstrate that you are paying attention to, and are interested in, others
- Notice your judgements about yourself and about others and challenge yourself to consider what they are based on

The self-reflection below is intended to develop your authenticity by helping you clearly define who you are, as a leader and as a human.
What are your most deeply held values (e.g. trust, truth, equality, community)?

....

....

....

....

How do your values influence your day-to-day actions?

....

....

....

....

What does being authentic mean to you?

....

....

....

....

How different is your persona as a leader from your personal persona? Which parts of yourself are you masking and what is causing you to do that?

....

....

....

....

Logic

Logic requires consistency between words and actions, and that your arguments or perspectives make sense. When people can trust your judgement and see that you are predictable (i.e. you do what you say you will), they are more likely to trust you. Some practices that can help here include:

- Explain your actions and decisions. Include motivations and values, to help others understand your reasoning
- Regulate and articulate your emotions. This helps in two ways. Firstly, it demonstrates that your decisions are thought-through and rational, not reactive
- Secondly, it ensures that you are consistent in your thinking, not erratic
- Demonstrate ethical behaviour in your interactions, such as sensitive handling of confidential information (e.g. neurodivergence diagnosis for a team member)

Empathy

Empathy is the "ability to understand and share the feelings of another".[22] It can help strengthen a group's collective identity and shared experience, reduce unproductive conflict, and improve communication.[23] See Table 6.1 for some examples of what empathy looks like.

Table 6.1 Examples of empathy

	Empathetic	*Lack of empathy*
Statement	"I've been feeling really overwhelmed with work lately."	
Response	"I'm sorry to hear that. It sounds tough. Is there anything specific that's causing you overwhelm? What can I do to help?"	"Well, everyone's really overwhelmed with work. There's nothing we can do about that."
Statement	"I've been having a lot of headaches recently, and I'm worried it might be something serious."	
Response	"That sounds concerning. Have you spoken to a doctor? I'm here for you and please let me know if it's impacting your work."	"It's probably nothing: most of these kinds of situations are nothing. Try drinking more water."

For some people, empathetic responses are automatic. For example, in one study, images of pain were shown to participants and they were asked to imagine pain in their own bodies. When they did this, the pain centres in the brain were activated; a situation such as this in which you feel pain as a result of someone else's pain is a demonstration of empathy.[24] However, for others, developing empathetic responses may take time and practice. Exercises that can help include asking open questions about others' experiences and proactively imagining yourself in another's situation.

Summary

To summarise, here is a list of indicators of personal trustworthiness as a leader:

1. Following through on your commitments
2. Admitting mistakes
3. Transparency around decisions and changes (where possible)
4. Giving credit to team members on merit
5. Actively listening and exploring others' ideas
6. Respecting confidential and sensitive information
7. Communicating expectations clearly and consistently
8. Handling conflicts and differences with openness and honesty

Which of these eight areas do you feel particularly good at?

....

....

....

....

In which situations do you exhibit less trustworthy behaviour?

....

....

....

....

What behaviours might you want to develop to increase your level of trustworthiness and in which contexts?

....

....

....

....

Case study: Safeguarding our own trustworthiness[25]

Presence in Action (PIA) Collective has been a community interest company (CIC) since 2019, with an incrementally growing membership of around 30 fee-paying members. It provides development and learning opportunities for member practitioners and the public using a complexity-attuned framework which all the members use as the primary resource to support themselves. The members belong to a range of professions, roles, sectors, ages and beliefs, representing a diversity of skills, experiences and thought. It is a self-organising community with no power hierarchy, in that the directors only take on an administrative governance obligation. Everyone who wishes to, takes part in decision-making about the organisation's operation, and all perspectives are welcomed and considered of equal value.

This psychologically safe context is first accessed through a four-day training programme (Power of Presence in Action). This opens participants up to further member-supported buddying practice and learning opportunities available through PIA Collective. Participants are introduced to seven community principles, and are invited to experiment with them in support of safeguarding their own trustworthiness in their own lives.

The primary commitment each member takes on is one of radical personal responsibility – as a way of attending to their impact on others and the wider world. Being amongst others embracing the same commitment generates an extraordinary level of co-created psychological safety. Witnessed by others, individuals review their engagement with their commitments annually. Those wishing to remain members ceremonially sign and submit a Principles of My Practice document which is held by the member organisation.

Non-judgemental reflections are encouraged in every PIA encounter. This means an individual may reflect what they notice occurring; they can share what they are feeling and what is coming up for them, seeking always to stay in relationship. They build on these insights to understand the personal implications for themselves and others. This practice helps everyone to learn about their own blind spots and biases, further reinforcing the culture of learning and radical responsibility. Everyone is resourced and encouraged to advocate on their own behalf, i.e. to speak up for themselves and to act with coherence in following through on their own and PIA Collective's commitments to nurture generative relationships.

An indication of the presence of a culture of high psychological safety is that members request and offer peer support for personal or work issues as and when required. Members also contribute new ideas or offerings which are discussed, after which actions are communicated. Further, members take responsibility and share that they have made mistakes, or work through their feelings, including embarrassment and fear. Individuals who have not been able to authentically be themselves in other groups have found themselves accepted for their difference, which includes neurodivergent thinking and behaviours.

Key points from Chapter 6

1. Leaders play an important role, through their behaviours, in creating an inclusive team environment.
2. The 6 Cs of inclusive leadership-commitment, cultural intelligence, courage, cognisance of bias, curiosity and collaboration - can create conditions for the whole team to thrive.
3. Psychological safety is influenced by leadership behaviours. Creating an environment in which people can bring their ideas, experiences, perspectives and personality to work makes more out of teams.
4. Creating a psychologically safe team culture takes time and practice. It starts with your own self-awareness and engaging in practices within your team.
5. Building trust among people is critical to psychological safety. Achieve this by being authentic, demonstrating logic in your decision-making, and being empathetic.

Notes

1 Bourke, J., "The six signature traits of inclusive leadership: Thriving in a diverse new world". *Deloitte insights*, 14 April 2016. Available at: https://www2.deloitte.com/us/en/insights/topics/talent/six-signature-traits-of-inclusive-leadership.html

2 Group-think is a phenomenon whereby the desire for harmony or conformity in a group leads to irrational or dysfunctional decision-making. Group members are likely to suppress dissenting opinions, prioritise consensus and overlook alternative perspectives, which can result in poor-quality decisions.

3 Edmondson, A.C., "Psychological safety and learning behaviour in work teams". *Administrative Science Quarterly*, 44(2) (1999), 350–83. Available at: https://www.researchgate.net/publication/313250589_Psychological_safety_and_learning_behavior_in_teams
4 Nembhard, I.M., and Edmondson, A., "Making it safe: The effects of leader inclusiveness and professional status on psychological safety and improvement efforts in health care teams". *Journal of Organisational Behaviour*, 27(7) (21 September 2006), 941–66. https://doi.org/10.1002/job.413.
5 Edmondson, A., *The Fearless Organisation: Creating Psychological Safety in the Workplace for Learning, Innovation and Growth*. New York: Wiley, 2018.
6 Gallo, A., "What is psychological safety?" *Harvard Business Review*, 15 February 2023. Available at: https://hbr.org/2023/02/what-is-psychological-safety
7 Gallo, A., "What is psychological safety?" *Harvard Business Review*, 15 February 2023. Available at: https://hbr.org/2023/02/what-is-psychological-safety
8 Kumar, S., "Psychological safety: What it is, why teams need it and how to make it flourish". *Chest Journal*, 165(4) (2024), 942–49. https://doi.org/10.1016/j.chest.2023.11.016. Edmondson, A., *The Fearless Organisation: Creating Psychological Safety in the Workplace for Learning, Innovation and Growth*. New York: Wiley, 2018.
9 Gallo, A., "What is psychological safety?" *Harvard Business Review*, 15 February 2023. Available at: https://hbr.org/2023/02/what-is-psychological-safety
10 Gallo, A., "What is psychological safety?" *Harvard Business Review*, 15 February 2023. Available at: https://hbr.org/2023/02/what-is-psychological-safety
11 Edmondson, A.C., "Psychological safety and learning behaviour in work teams". *Administrative Science Quarterly*, 44(2) (1999), 350–83. Available at: https://www.researchgate.net/publication/313250589_Psychological_safety_and_learning_behavior_in_teams. Bourke, J., and Titus, A., "Why inclusive leaders are good for organizations, and how to become one". *Harvard Business Review*, 29 March 2019. Available at: https://hbr.org/2019/03/why-inclusive-leaders-are-good-for-organizations-and-how-to-become-one.
12 Gallo, A., "What is psychological safety?" *Harvard Business Review*, 15 February 2023. Available at: https://hbr.org/2023/02/what-is-psychological-safety
13 Clarke, T., *The Four Stages of Psychological Safety: Defining the Path to Inclusion and Innovation*. San Francisco: Berrett-Koehler, 2020.
14 Cappelli, P., and Eldor, L., "Can workplaces have too much psychological safety?" *Harvard Business Review*, 3 January 2024. Available at: https://hbr.org/2024/01/can-workplaces-have-too-much-psychological-safety
15 Cappelli, P., and Eldor, L., "Can workplaces have too much psychological safety?" *Harvard Business Review*, 3 January 2024. Available at: https://hbr.org/2024/01/can-workplaces-have-too-much-psychological-safety
16 Hoover, D.W., "The effects of psychological trauma on children with autism spectrum disorders: A research review". *Review Journal of Autism and Developmental Disorders*, 2 (9 July 2015), 287–99. https://doi.org/10.1007/s40489-015-0052-y. Lobregt-van Burren, E., Hoekert, M., and Sizoo, B., "Autism, adverse events and trauma". In Grabrucker, A.M., *Autism Spectrum Disorders*. Netherlands: Exon Publications, 2021. Szymanski, K., Sapanski, L., and Conway, F.,

"Trauma and ADHD-association or diagnostic confusion? A clinical perspective". *Journal of Infant, Child, and Adolescent Psychotherapy*, 10(1) (2011), 51–59. https://doi.org/10.1080/15289168.2011.575704. Alexander-Passe, N., "Investigating Post Traumatic Stress Disorder (PTSD) triggered by the experience of dyslexia in mainstream school education?" *Journal of Psychology and Psychotherapy*, 5(6) (January 2015). https://doi.org/10.4172/2161-0487.1000215. Wadsworth, L.P., Van Kirk, N., August, M., *et al.*, "Understanding the overlap between OCD and trauma: Development of the OCD trauma timeline interview (OTTI) for clinical settings". *Current Psychology*, 42 (2023), 6937–47. https://doi.org/10.1007/s12144-021-02118-3. Van der Kolk, B.A, "Clinical implications of neuroscience research in PTSD". In *Annals of the New York Academy of Sciences*, 1,071(1), *Psychobiology of Posttraumatic Stress Disorder: A Decade of Progress*, July 2006. Van der Kolk, B.A., "The neurobiology of childhood trauma and abuse". *Child and Adolescent Psychiatric Clinics*, 12(2) (2003), 293–317. https://doi.org/10.1016/s1056-4993(03)00003-8. Van der Kolk, B.A., "Posttraumatic therapy in the age of neuroscience". *Psychoanalytic Dialogues*, 12(3) (2002), 381–92. https://doi.org/10.1080/10481881209348674

17 Iqbal, Q., Volpone, S.D., and Piwowar-Sulej, K., "Workforce neurodiversity and workplace avoidance behaviour: The role of inclusive leadership, relational energy, and self-control demands". *Journal of Human Resource Management*, 64(1) (1 August 2024), 37–57. https://doi.org/10.1002/hrm.22249

18 Edmondson, A. and Hugander, P., "4 steps to boost psychological safety at your workplace". *Harvard Business Review*, 22 June 2021. Available at: https://hbr.org/2021/06/4-steps-to-boost-psychological-safety-at-your-workplace

19 Frei, F., and Morriss, A., "Begin with trust: The first step to becoming a genuinely empowering leader". *Harvard Business Review*, May/June 2020. Available at: https://hbr.org/2020/05/begin-with-trust?registration=success

20 Brown, B., *Daring Greatly: How the Courage to be Vulnerable Transforms the Way we Live, Love, Parent and Lead*. New York: Avery, 2015.

21 Korsgaard, M.A., "Reciprocal trust: A self-reinforcing dynamic process". In Searle, R.H., Nienaber, A.-M.I., and Sitkin, S.B. (eds), *The Routledge Companion to Trust* (pp.14–28). Abingdon, UK: Routledge, 2018.

22 Fletcher-Watson, S., and Bird, G. "Autism and empathy: What are the real links?" *Autism*, 24(1) (2019), 3–6. https://doi.org/10.1177/1362361319883506

23 Klimecki, O.M., "The role of empathy and compassion in conflict resolution". *Emotion Review*, 11(4) (2019), 310–25. https://doi.org/10.1177/1754073919838609

24 Ogino, Y., Nemoto, H., Inui, K., Saito, S., Kakigi, R. and Goto, F., "Inner experience of pain: Imagination of pain while reviewing images showing painful events forms subjective pain representation in humans' brain," *Cerebral Cortex*, 17(5) (2007), 1139–46. https://doi.org/10.1093/cercor/bhl023

25 Gardiner, L.J., 2019. Attending, daring, becoming: Making boundary-play conscious. In J. Birch and P. Welch (eds), *Coaching Supervision* (pp. 103–125). Abingdon, UK: Routledge. Gardiner, L.J.N., *Attending Responding Becoming: A Living-Learning Inquiry in a Naturally Inclusional Playspace*. Doctoral thesis, University of Hull, 2022.

7 Taking a strengths-based approach

As we have seen in previous chapters, neurodivergent individuals have unique strengths that they can contribute to their workplaces and their teams. However, a majority of people – of all types – find that their strengths aren't being harnessed effectively at work: one survey shows that only a third of people are given the opportunity to do what they are best at on a daily basis.[1]

Think about your last performance review. What part of the review did you spend most time on? My guess is that you spent more time talking about your development areas than you did about how you could use more of your strengths. There are two main reasons behind this. Firstly, we are encouraged from a very young age to focus on our development areas rather than our strengths:[2] school reports, for example, would focus on what wasn't done well – the basic learning model of the UK education system. Secondly, intrinsic negativity bias draws us towards negative data in favour of positive data,[3] again leading us to look at fixing what's not working.

This is clearly a missed opportunity. Making more of our strengths has also been shown to improve quality of life, increase engagement at work and elevate levels of motivation.[4] Studies have also shown that matching neurodivergent individuals' strength profiles (knowledge, skills and abilities as well as social requirements) to the role can promote successful employment.[5] Neurodivergent individuals have shown a preference for these kinds of approaches because they are more motivational than the traditional "medical model" deficit approach (see Chapter 1).[6]

DOI: 10.4324/9781003645535-7

When were you at your best at work?

Let's start with another self-reflection exercise. Think about a time when you felt your happiest and most engaged at work.

What role were you doing?

....

....

....

....

What tasks did that entail?

....

....

....

....

Which of your natural talents were you using?

....

....

....

....

Did others acknowledge and appreciate your natural talents?

....

....

....

....

According to the strengths-based methodology, if you are describing a time when you were happy at work, it is likely one in which you were using your natural talents.

What is a strengths-based approach?

This approach proposes that we are happiest, most fulfilled and therefore more successful when we are able to utilise our strengths, rather than focusing on things we are less able to do.

Everyone has *natural talents*. These are the things that we are drawn to, that give us energy, which we naturally gravitate towards and which we find easier. We gain *experience*, which is the repetition of a task or the use of capabilities that enable us to build a positive association with the task and to get better at it. And then we have *skills*, which is the knowledge or expertise associated with a subject. It might be what you learn at school or university, or the technical knowledge required in your role. When you have the opportunity to develop your natural talents through skills and experience, you find that you are working in an area of strength for you. It's when we are using our strengths that we are at our best.

Think about a racing car driver. Exceptional racing car drivers are born with quick reaction times, a risk-taking mindset and spatial awareness; these are their natural talents. They develop these talents by learning about the car and how it works (skills) and repeatedly driving round the track (experience). The combination of these three things is what gives them a strength as a racing car driver.

But if you were born without those natural talents - reaction times, mindset and spatial awareness - then no matter how many times you drove round the track or how much you learnt about the car, you would likely never be as good as someone who had those natural talents. You may well increase your capability, but driving would probably deplete your energy as you would have to work so hard at it.

The same goes for the workplace. If we are naturally talented at a task, we will gravitate towards it and get energy from it. Absent that natural talent, it will take more energy to perform it and we may find it exhausting. If we have to do it - possibly driven by social norms, i.e. in order to fit in - it can become a learnt behaviour, performed well through experience and skills but not energising.

Let's look at this in relation to neurodivergent behaviours and patterns:

- **Organisation**. Not a natural talent for some ADHD or dyslexic individuals. They might have been given the tools (skills) to stay on top of their work (e.g. to-do lists, time management techniques and apps to support prioritisation), and they may have the opportunity to do these tasks day in day out (experience); but they will likely always find doing so takes more effort than it would for those with a natural talent for organisation. However, it can become a learnt behaviour
- **Detail orientation**. A natural talent for some autistic individuals, who can be detail-oriented and able to see patterns in datasets. This brings them energy and they will gravitate towards it. Being in a role in which they can develop this skill through training, and use these skills repeatedly (experience), enables them to capitalise on this talent and feel more confident. They may also feel more valuable to the team, and this may reduce the urge to compare themselves less favourably against other team members

This approach also underscores the importance of having people assigned to the right roles and having the roles crafted around individuals. When people are in roles that play to their natural talents, they will find the work less tiring, more fulfilling and, ultimately, they will be able to develop. Expending extra energy on tasks using learnt behaviours could contribute to neurodivergent individuals' stress and lead to burnout. Furthermore, doing more of what makes you happy and confident, i.e. using strengths, can result in improved self-efficacy (your ability to perform a task). An improvement in self-efficacy is positively associated with wellbeing and quality of life.

In a neurotypical world, neurodivergent individuals are often not given the opportunity to use their natural talents as much as their neurotypical colleagues. The importance of this approach is therefore clear.[7]

An approach to mapping strengths

Neurodivergent traits can be both strengths and challenges. ND individuals can have exceptional cognitive abilities in some situations but notably poor ones in others. Drawing out these cognitive abilities to create a profile can help raise self-awareness, highlight strengths and aid discussions between managers and their teams about capabilities and gaps in requirements. This approach - rating capabilities across a range of areas, then mapping them on a chart - results in what is known as a "spiky profile".[8]

Figure 7.1 shows two spiky profiles mapped on the same chart. Ratings go from 0 (an area of difficulty) to 5 (an area of strength). The dark line is that of a neurotypical individual. Notice the small range: every score falls between 3.2 and 4.2. The lighter line is that of an autistic individual. Notice the larger range of scores, ranging from 1 out of 5 to 4.9: this is what makes the profiles "spiky". The implication of this is that neurodivergent individuals can make better specialists than generalists. We must be careful with this generalisation, however: it depends on how the areas of strength map on to potential roles, and every individual is different.

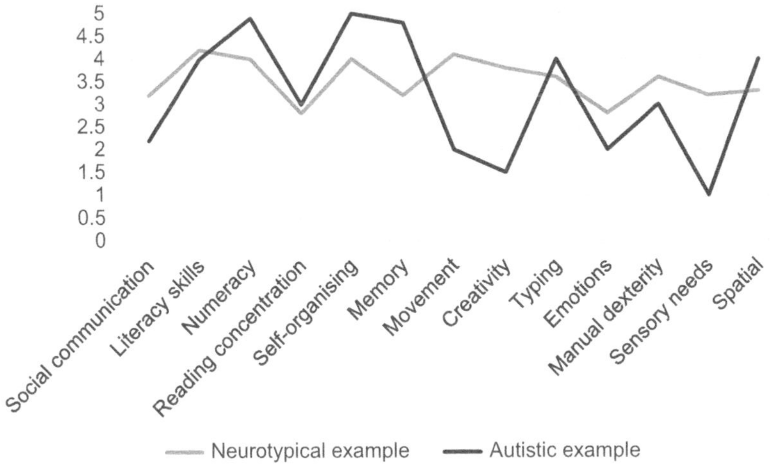

Figure 7.1 Spiky profiles: neurotypical and autistic examples

Note: Diagram based on research by Doyle, N. (2017)[9]

Strength	% reporting this strength
Hyperfocus	80.0
Creativity	78.1
Innovative thinking	75.0
Detail processing	71.0
Authenticity	64.4
Visual reasoning	58.3
Long-term memory	55.3
Entrepreneurialism	46.0
Verbal comprehension	44.1
Cognitive control	40.1
Visual-spatial skills	33.4
Numeracy	30.1
Short-term memory	14.8

Figure 7.2 Strengths reported by neurodivergent individuals

Source: McDowall, A., Doyle, N. and Kiseleva, M., Neurodiversity at Work 2023. University of London, Birkbeck, 2022. Neurodiversity in Business Report commission.

Under a strengths-based approach, you would work on how the strengths could be used more productively. For example, if someone, like the person in Figure 7.1, is great at numeracy, could they do more analysis-based tasks? Or, if literacy is their strength, what tasks might involve summarising data in words for presentation to others?

Based on research from Birkbeck at the University of London,[10] Figure 7.2 presents strengths as self-reported by neurodivergent individuals.

It is possible to overuse strengths:

- Hyperfocus, the top strength listed in Figure 7.2, is a capability typically linked to ADHD and autism. It allows individuals to focus intensely on a task, which is seen to sometimes result in increased productivity.[11] However, hyperfocus can also lead to a dereliction of other responsibilities, e.g. forgetting to go to the toilet or to eat, or it can result in burnout[12] and consequently an unwillingness to return to that kind of project. Managers and leaders need to have conversations about using people's strengths in context: for example, an individual with hyperfocus might need a reminder to take a break
- Creativity and innovation, the second strength in Figure 7.2, can also be overused. Imagine someone in a problem-solving discussion who keeps throwing ideas out long after the time when the

discussion needed to be wrapped up. Or perhaps they are so busy thinking of creative solutions that they miss the other part of their role: to turn some of those ideas into action plans. In these situations, raise awareness of the overuse, and partner with the individual in question to discuss whether the role needs to change to accommodate their strength, or if other approaches could prevent them from overusing their strength

What are *your* strengths?

There are many profiling tools to identify your strengths and natural talents: for example, the (free) VIA Character Strengths Survey, or the Clifton StrengthsFinder. But you can start here by reflecting on the ten questions below. Answer them from gut-feeling if possible, for the most accurate outcome. If you are struggling with the language needed to describe strengths, take your cue from the list In Table 7.1. It's not an exhaustive list but rather a starting point for inspiration.

Table 7.1 84 words to describe strengths

Creative	Compassionate	Ambitious	Thorough
Analytical	Perceptive	Empowering	Self-disciplined
Empathetic	Original	Humorous	Balanced
Innovative	Diligent	Articulate	Humble
Resilient	Problem-solver	Tactful	Mindful
Detail-oriented	Enthusiastic	Visionary	Self-sufficient
Adaptable	Optimistic	Motivated	Persistent
Resourceful	Goal-oriented	Systematic	Rational
Determined	Thoughtful	Fair-minded	Altruistic
Strategic	Artistic	Outgoing	Intuitive
Focused	Honest	Socially aware	Diplomatic
Logical	Practical	Supportive	Open-minded
Insightful	Conscientious	Warm	Steadfast
Curious	Reliable	Charismatic	Courteous
Imaginative	Inquisitive	Pragmatic	Grounded
Methodical	Collaborative	Candid	Generous
Independent	Proactive	Driven	Meticulous
Tenacious	Flexible	Trustworthy	Passionate
Versatile	Efficient	Loyal	Patient
Inventive	Calm	Energetic	Philosophical
Precise	Ethical	Observant	Pioneering

What's your favourite part of your job? What are you doing when you are happiest at work?
I love it when...

....

....

....

....

What do you find easy that others don't?

....

....

....

....

Can you remember a time when you felt really good about using your strengths at work?

....

....

....

....

What have you been recognised for doing well – something you could keep doing for a long period of time because it gives you energy?

....

....

....

....

What activities are always fun for you?

....

....

....

....

On days when you are looking forward to work, what exactly are you looking forward to doing?

....

....

....

....

What kinds of activities are you able to learn more quickly?

....

....

....

....

Practices to empower your team using strengths

Identify

- Work with your team member to identify their strengths. Give some consideration to those strengths conferred by different ways of thinking. Tell them how you see these strengths showing up in their work, and their benefits for the team
- Identifying strengths can take more time and effort depending on the person. This is particularly true for neurodivergent individuals who have prior experience of their capabilities not being recognised. At school, commonly, they might have been called "stupid" for not being able to read or write well, or "disruptive" for having a lot of energy in the classroom. Changing the patterns of thinking embedded by these experiences takes time and practice. Managers can support this by coaching them through the process of identifying strengths, and, having identified them, acknowledging and celebrating them (see "Acknowledge and celebrate" below).

Ask others what they see as the team member's strengths and allow the team member to hear these responses first-hand. Shape the conversations with questions such as these:

- What characteristics make me good to have on a team?
- When do you see me operating at my best?
- What unique value do I bring to the team?
- What three words would you use to describe me?

Acknowledge and celebrate

- Find ways to acknowledge their strengths day to day, tailoring your approach to the individual. For example, some would appreciate having their strengths called out publicly; others would feel embarrassed, so you would celebrate privately instead

Adjust

- Partner with your team members to tailor their roles so that they can lean into their strengths. For example, are there more creative tasks that could leverage their innovative mindset? Is there detail-oriented work to use their ability to spot errors? Could you partner them with someone else so that they can both play to their strengths and mitigate each other's development areas?

A strengths-based approach can improve the whole team's performance.[13] So review your team's overall strengths to see which ones you are missing. Then you can tailor roles or hire into those gaps, as well as allocating work in line with the capabilities already there. The exercise below supports this kind of analysis.

Exercise: What are your team's strengths?

See Table 7.2. Complete it for your team members. If you are struggling to find the language to describe their strengths, reflect on the questions in the reflection exercise above ("What are *your* strengths?"), or use the list of terms in Table 7.1.

Table 7.2 What are your team's strengths?

Team member name	*Strengths they have*	*How much do I make use of these strengths? (scale of 1–10 where 0 = not at all, 10 = fully utilised)*	*What could I do to use this strength more?*

How to manage development areas

The decision tree shown in Figure 7.3 helps you to determine the approach for each development area. Let's look at an example of a team member who has been told that, in order to get promoted, they need to network with senior leaders more. The underlying issue here might be one of the following:

- They are missing the *natural talent* of reading social situations
- They might not have had *experience* in building connections with others
- They haven't been shown, or had role-modelled, what it would look like to network effectively (*skills*)
- They may not feel safe enough to network without fear of judgement or failure (*lack of psychological safety*)
- They have *overused* their talent

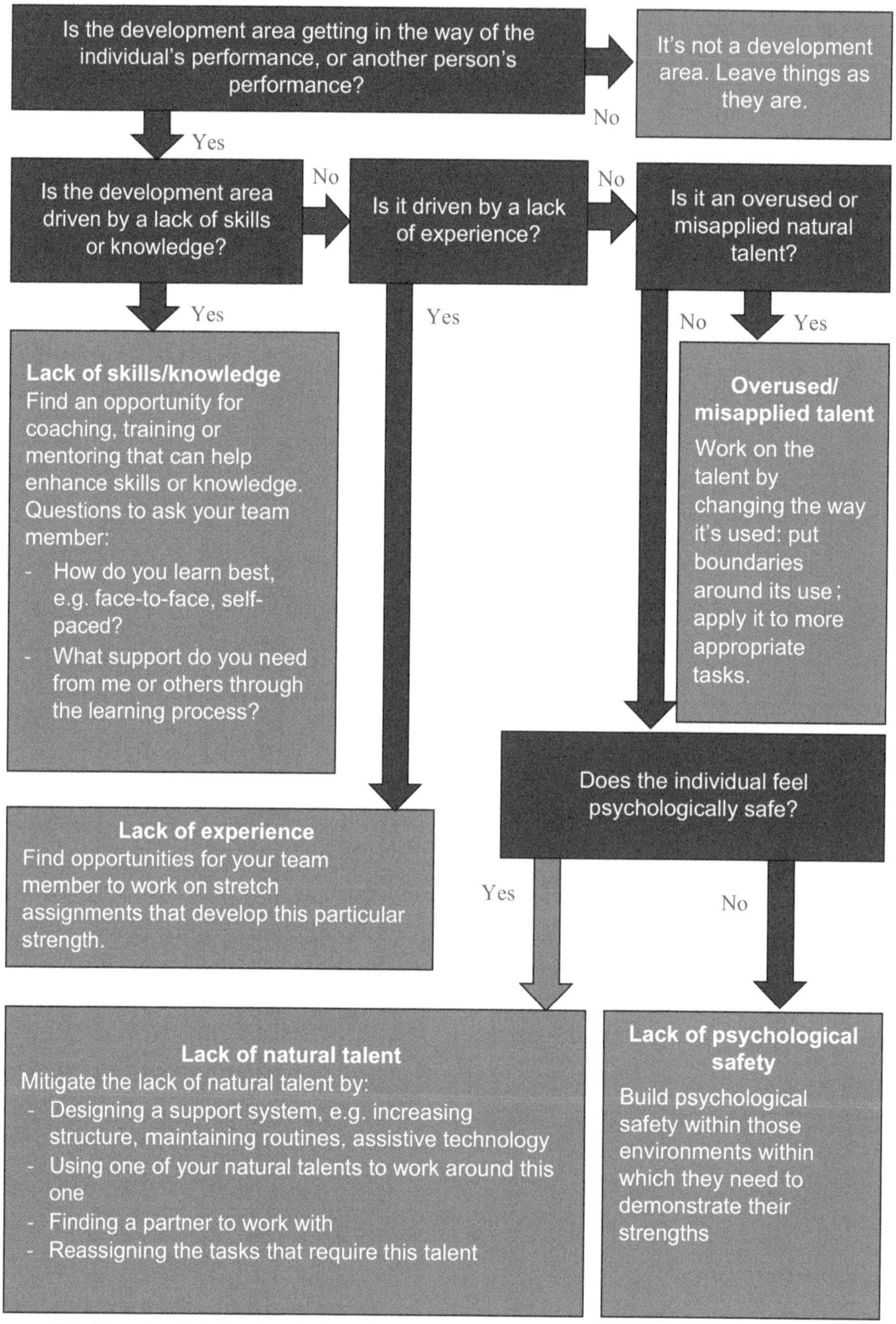

Figure 7.3 How to think about development areas

Having identified the development-area category, you can go about managing it. For example:

- If they are missing the natural talent of reading social situations, you could pair them with someone who is naturally good at it, who can help them work through such situations. Alternatively, provide assistive technology, or strategies that use their strengths in order to mitigate their development areas[14]
- If they haven't had experience in building connections, can they be given opportunities to practise with a low-risk audience (i.e. where promotion is not attached to the outcome)?
- If they don't know what effective networking looks like, they could be assigned a mentor or choose to go on a training course
- If they don't feel safe enough to demonstrate their strengths, how can you, as their manager, put them in situations where they do feel safe enough? (See Chapter 6)

Case study: Using strengths-based assessments

I completed what was known then as Realise 2 CAPP strengths[15] as part of my development review. I agreed with the analysis except for two areas which was the learnt behaviour of "work ethic".

I was at a loss about my work ethic being a learnt behaviour. I was conscientious and hard-working. I enjoyed my job most of the time. It was the world to me at this time of my life. I was the youngest manager in our team; I had recently completed an MBA and two secondments. What else did I need to do to prove my work ethic? So I doubted this result of the assessment and wondered if it was the way I answered the questions. I had interpreted the results as them telling me I wasn't really hard-working enough and I was defensive about this.

I felt angry with the review and I felt disappointed with the results. But all of the results except this one but made sense. It clashed with my values about being hard-working. So I explored this with a coach who I had been assigned to for career coaching. Through this exploration I began to realise this theme was far bigger than I had first thought.

We uncovered one reason why I was so "hard-working". I had put in long hours, working 12-hour days and often travelling over weekends, trying to prove I was good at my job. But why? Surely I had done that with my promotions and how my team performed? I still doubted my worth.

Our company culture was fast-moving and we were constantly restructuring. I had learnt if you weren't of value you were out! But this wasn't just about the culture of the company but also about why I thought I wasn't good enough. I noticed I was very judgemental about other people who didn't in my view "work hard". I thought they were lazy and if they worked for me my judgement of them tarnished everything they did. I didn't see the value or skill I just saw that they didn't try hard enough.

The thing that made me accept this was a perception about me was the fact that I felt exhausted. I was always tired and had little energy for anything else in my life except work. So my work ethic couldn't be a strength if I was so tired and if it impaired my judgement of others.

When I looked far enough back, I realised because school had been so difficult I was still trying to prove myself. I remembered teachers telling me I didn't work hard enough because I was distracted and found it difficult to write. I was believing everyone had to "bust a gut" like me to be worth it because of my low self-esteem. I had to work harder than others to read and write reports because it took longer. With more admin in a managerial position my work ethic was becoming unhealthy for me. It was a revelation I wouldn't have picked up if it wasn't for the strengths exercise and someone to talk through the results.

30-year-old female working in Biotech who has ADHD and dyslexia

Case study: Use of strengths-based approaches

Siemens is a European-based manufacturing organisation with over 350,000 employees. The nature of its business means that it is reliant on a steady stream of entry-level talent (interns, apprentices and graduates), who develop their skills to become engineers. However, as many job seekers approaching Siemens for a role have minimal work experience, using a standard interview technique that requires candidates to draw on their own experiences can result in rehearsed answers and limitations to the answers they could give.

In 2017, the company shifted its approach away from ability and towards mindset and behaviours - looking particularly at potential, passions and energy. The different job roles were analysed with a focus on the strengths that would identify high and low performers. This resulted in a strengths-based interview script and assigned training for managers[16].

It worked. Recruiters and hiring managers said they had more data on the candidates and therefore more insight into what they might be capable of. Hiring decisions were easier, and the candidates spoke of a better interview experience in which they were able to demonstrate their full potential. Of the managers, 82% agreed or strongly agreed that strengths-based interviews were fair, consistent and objective - up from 58% for the competency-based process.

Key points from Chapter 7

1. When people are working in areas that build on their natural talents, they are likely to be more energised, confident and capable. Matching the strengths of neurodivergent individuals to their role has been shown to enable career success.
2. However, workplaces often focus on mitigating areas of development rather than focusing on strengths. This can mean that

people are working in areas that come less naturally to them and use more of their energy. For some neurodivergent individuals, this is linked to burnout and stress.

3. A "spiky" profile can be used to identify a neurodivergent person's strengths and challenges. This can support a discussion about how they can best use their talents at work. (Other positive psychology strengths-based assessments are available online, such as the VIA Character Strengths Survey).
4. Development areas may arise as a result of overused talents, lack of skills, insufficient experience or a deficit of natural talents in that area. Identifying what drives a development area helps you find the right way to tackle it.
5. Reviewing the strengths of your team overall informs your people development strategy and supports resource allocation processes.

Notes

1 Clifton, D., *Strengthsfinder 2.0 from Gallup and Tom Rath: Discover your CliftonStrengths*. London: Gallup Press, 2017. Buckingham, M., and Clifton, D., *Now, Discover Your Strengths: How to Develop Your Talents and Those of the People You Manage*. New York: Simon & Schuster, 2004. Schutte, N.S., and Malouff, J.M., "The impact of signature character strengths interventions: A meta-analysis". *Journal of Happiness Studies: An Interdisciplinary Forum on Subjective Well-Being*, 20(4) (2019), 1179-96. https://doi.org/10.1007/s10902-018-9990-2

2 Buckingham, M., and Clifton, D., *Now, Discover Your Strengths: How to Develop Your Talents and Those of the People You Manage*. New York: Simon & Schuster, 2004.

3 Rozin, P., and Royzman, E.B., "Negativity bias, negativity dominance, and contagion". *Personality and Social Psychology Review*, 5(4) (2001), 296-320. https://doi.org/10.1207/S15327957PSPR0504_2

4 Clifton, D., *Strengthsfinder 2.0 from Gallup and Tom Rath: Discover Your CliftonStrengths*. London: Gallup Press, 2017. Buckingham, M., and Clifton, D., *Now, Discover Your Strengths: How to Develop Your Talents and Those of the People You Manage*. New York: Simon & Schuster, 2004.

5 Bölte, S., Carpini, J.A., Black, M.H., Toomingas, A., Jansson, F., Marschik, P.B., Girdler, S. and Jonsson, M., "Career guidance and employment issues for

neurodivergent individuals: A scoping review and stakeholder consultation". *Journal of Human Resource Management*, 64(1) (28 October 2024), 201–77. https://doi.org/10.1002/hrm.22259

6 Schrevel, S.J.C., Dedding, C. and Broerse, J.E.W., "Why do adults with ADHD choose strength-based coaching over public mental health care? A qualitative case study from the Netherlands". *Sage Open*, July–September 2016, 1–8. https://doi.org/10.1177/2158244016662498

7 Black, M.H., Helander, J., Segers, J., Ingard, C., Bervoets, J., Grimaldi de Puget, V. and Bölte, S., "Resilience in the face of neurodivergence: A scoping review of resilience and factors promoting positive outcomes". *Clinical Psychology Review*, 113 (November 2024), 102487. https://doi.org/10.1016/j.cpr.2024.102487

8 Doyle, N., "Neurodiversity at work". In British Psychological Society (ed.), *Psychology at Work: Improving Wellbeing and Productivity in the Workplace*, 44–62. Leicester, UK: British Psychological Society, 2017. Grant, D., "The psychological assessment of neurodiversity". In D. Pollak (ed.), *Neurodiversity in Higher Education*, 33–62. Chichester, UK: Wiley-Blackwell, 2009.

9 Doyle, N.,"Neurodiversity at work". In *Psychology at Work: Improving Wellbeing and Productivity in the Workplace*, ed. British Psychological Society, 44–62. Leicester, UK: British Psychological Society, 2017.

10 McDowall, A., Doyle, N. and Kiseleva, M., *Neurodiversity at Work 2023*. University of London, Birkbeck, 2022. Neurodiversity in Business Report commission. Note that there was a disproportionate number of autistic and ADHD individuals within the neurodivergent population that they sampled, so there may be a bias towards traits of those neurotypes. Allowing for that, the results demonstrate the breadth and uniqueness of the strengths being brought to the workplace by neurodivergent individuals.

11 Saline, S., "Hyperfocus and ADHD: Productivity superpower or Kryptonite?" *Psychology Today*, 22 March 2022. Available at: https://www.psychologytoday.com/gb/blog/your-way-adhd/202203/hyperfocus-and-adhd-productivity-superpower-or-kryptonite. Ashinoff, B.K., and Abu-Akel, A., "Hyperfocus: The forgotten frontier of attention". *Psychological Research*, 85(1) (2021), 1–19. https://doi.org/10.1007/s00426-019-01245-8

12 Adamou, M., Arif, M., Asherson, P., *et al.*, "Occupational issues of adults with ADHD". *BMC Psychiatry*, 13(1) (2013), 59. https://doi.org/10.1186/1471-244X-13-59

13 Clifton, D., *Strengthsfinder 2.0 from Gallup and Tom Rath: Discover Your CliftonStrengths*. London: Gallup Press, 2017. Buckingham, M., and Clifton, D., *Now, Discover Your Strengths: How to Develop Your Talents and Those of the People You Manage*. New York: Simon & Schuster, 2004.

14 An interesting study has shown that chatbots can be used to apply a strengths-based solution to development areas for neurodivergent workers. Hall, K., Arora, P., Lowy, R. and Kim, J.G., "Designing for strengths: Opportunities

to support neurodiversity in the workplace". In *Proceedings of the 2024 CHI Conference on Human Factors in Computing Systems* (CHI '24). New York: Association for Computing Machinery, 2024. Article 74, 1-14. https://doi.org/10.1145/3613904.3642424
15 Linley, A., *Average to A+: Realising Strengths in Yourself and Others*. Coventry, UK: CAPP Press, 2008.
16 Strengthscope, "Case study: Siemens. Supercharging entry level recruitment at Siemens". Strengthscope, n.d. Available at: https://www.strengthscope.com/case-studies/siemens

8 Managing change

As Heraclitus allegedly said, change is the only constant.[1] And that is certainly true of modern-day workplaces, whether it be system change, process change or personnel change. Managing change is difficult for many, and can cause anxiety, doubt and insecurity. Yet it provides opportunities for personal growth, innovation of products and better use of resources.

Neurodivergent individuals may have a relationship to change that differs from what might be considered typical or usual. For example:

- Autistic individuals may have routines to help to manage their anxiety, which can be impacted by changes, e.g. a desk move in the office
- Some neurodivergent individuals struggle with emotional dysregulation (i.e. poor management of emotional responses) and/or alexithymia (i.e. difficultly in recognising, expressing and identifying emotions). This makes it more difficult to handle the emotional components of change[2]
- ADHD individuals can intentionally create change through disruption when they are bored. For example, having been asked to follow a process, they found it more motivating to change the process and took a different approach instead
- A dyslexic may find online retraining off the back of a role change difficult, as a result of finding it hard to follow the text

DOI: 10.4324/9781003645535-8

Case study: Communicating change

During the financial crisis of 2008, I went through a redundancy process. A matrix scoring system was used to decide who would be made redundant and the communications said this had been applied across the whole firm. I scored one point lower than the other director in my office, so I was the one to go. But I scored a lot higher than many of the directors in the London office and offered to move to London as my role should have been safe based on my score. I was then told each office was doing the process in isolation and so it wasn't an option. This felt very unfair, which is a huge trigger for my autism. I still feel triggered when I think of this. I understood the need for the redundancies as there wasn't enough work to sustain the corporate finance teams when the markets crashed, but I couldn't reconcile with the way it was handled.

49-year-old female senior manager in professional services

Often, we focus on the direct impact of the change, but the way in which the change affects personnel can be more indirect. For example, a technology system change (direct impact), which results in the assistive technology used by a dyslexic individual not working for them any more (secondary impact). Therefore, consider all the knock-on, or secondary, impacts as well as direct ones, and consult team members about them as early in the process as possible. For example, for public sector organisations in Scotland, consideration of these impacts would be done through an Equality Impact Assessment[3] as part of the equality duty.[4] This is a helpful framework for thinking through unintended consequences, particularly consequences that fall on those with disability and other protected characteristics (Chapter 2).

Everyone responds to change differently, and every situation is different. And there is limited research on how different neurotypes respond to workplace changes. This chapter therefore provides frameworks for navigating change applicable to all people and all situations with a specific focus on neurodivergent individuals. The key, as we have seen in previous chapters, is to remain curious and empathetic with your team members, particularly when their response is different to yours.

Your relationship to change

In the following reflection exercise, let's start by listing all the different kinds of changes that you are experiencing in your workplace right now, from small to large.

....

....

....

....

How do you feel about each of these changes? Anxious? Excited? Sad? Curious? Something else?

....

....

....

....

What patterns or trends do you notice about your emotions? For example, which kinds of changes trigger which kinds of emotions? What was it specifically about the changes and how they were implemented? How did that impact your emotional response?

....

....

....

....

Exploring our emotional reaction to change

Frequently overlooked by managers and leaders is the emotional impact associated with change. Research has shown that when emotions around a change are managed well, there is a direct positive impact on motivation, work quality and commitment to the change.[5]

The Kübler-Ross change curve®,[6] originally formulated to describe the stages of grief, has been widely adopted to understand and anticipate emotional responses to change in the workplace. The curve identifies a series of emotional stages that individuals typically experience when confronted with significant change in their lives (see Figure 8.1).

The Y axis represents level of energy; in some models, it represents self-esteem and morale.[7] Although the X axis represents time, that doesn't mean people move through stages 1, 2 and 3 in a linear way; you can go backwards and forwards up and down the curve several times. Nonetheless, identifying the stage that a team member is at informs how you might best support them at that time.

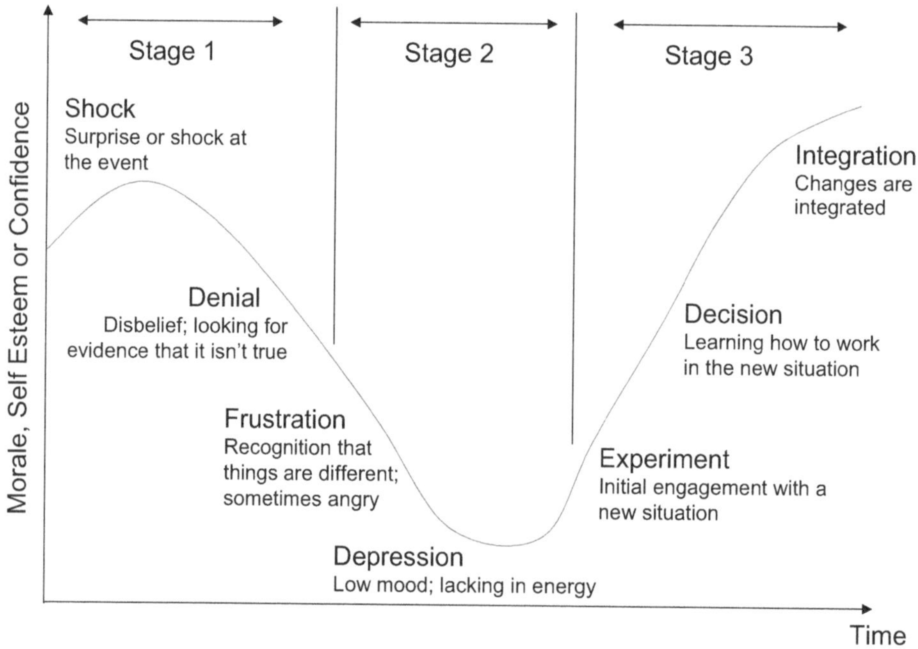

Figure 8.1 The Elisabeth Kübler-Ross (EKR) Change Curve®

Stage 1: Shock and denial

Employees have not yet processed the news of the change. Maybe they have just been told they are being made redundant, without any notice, and they are in shock. They may be trying to figure if they were the only ones that didn't see this coming, retracing their thoughts in disbelief. Here, consider an individual's previous experiences and how this might affect their response to change. Many neurodivergent individuals have had bad experiences with change, e.g., redundancy or poorly managed reorganisations, and this is likely to have an effect on how they respond to (an)other change of this type.

They may fall into denial and resist the plan, choosing instead to believe that it isn't really going to happen. This may manifest in an unwillingness to talk about what happens next, or a dismissive tone, or withdrawal from conversations. For example, a company announces a major technical upgrade and some employees continue using the old system, convincing themselves that the implementation will be abandoned and perhaps, with time, everyone else will realise this change isn't a good idea. Those in the denial stage may try to convince others that the change won't happen in order to reinforce their own denial. To help them through his stage, be empathetic and remember how you felt when you first heard about this news, or when you were in a similar situation and resisted a change.

Uncertainty and lack of control is difficult for some neurodivergent individuals, potentially resulting in obsessive or ritualistic behaviours[8] and emotional dysregulation. Provide as much certainty as you can to help alleviate this. Where possible, express the changes in terms of how they might impact daily tasks, and offer 1:1 conversations about adapting routines. Being supportive in this way provides reassurance. Tailor your communication to your audience (see Chapter 5): if you are moving office, for example, show pictures of the new office rather than describing it verbally, if your team member is a visual person. Refer to the changes in a non-personal way, i.e. not in relation to their performance or what they have or have not done.

Stage 2: Frustration/anger and depression

In stage 2 we start to experience the reality of the change. Now, employees may be angry, feeling frustrated and helpless, and direct their anger towards management or the change itself. They might attempt to regain control by bargaining the terms of the change; if they realise that bargaining won't work, they may feel depressed, not knowing what they can do. They may consequently suffer loss of sleep or appetite, or be under- or overactive compared to previous levels.[9] For example, a company decides to downsize, and the employees harbour resentment towards leadership, perceiving the decision as a betrayal. They express that as anger, start to query the terms of the redundancies and try to bargain. They ultimately feel down when they realise that the changes are inevitable and there are things they cannot control.

This is why acknowledging emotions associated with the change is so important. As we saw, neurodivergent individuals can experience alexithymia, emotional dysregulation and rejection sensitivity dysphoria. This can mean stronger emotional outbursts than you might expect, or increased intensities of other feelings such as shame or embarrassment. It is your role as a manager to provide spaces for individuals to discuss and process their emotions, and this includes neutral spaces with high psychological safety (Chapter 7), such as employee assistance programmes. Keep an eye out for impulsive reactions, e.g. directly complaining to senior management in a career-limiting way. Supporting people through this stage may require multiple reiterations of the rationale for change to ensure that it has been processed and understood.

Some neurodivergent individuals have a predilection towards depression and anxiety,[10] as explored in Chapter 2. They may experience greater feelings of depression than others in the team, or it may take them longer to progress through this stage. Options for professional support can be helpful, as can means of communication that don't require words: e.g. a colour code such as "red" if you are not OK, "yellow" for OK and "green" if you are doing great. Tools such as the blob tree, emotions wheel or Emotions Palette™ help individuals to understand where they are and to be more emotionally literate (see Chapter 12 for more on these resources).

Giving people agency and choices through the process increases wellbeing, increases feeling of calm and enhances performance.[11] Neurodivergent individuals may experience choice and decision-making differently to neurotypical individuals,[12] so providing a list of offerings can reduce overwhelm and increase feelings of autonomy for some.

Stage 3: Experimentation, acceptance and integration

In this stage, we start to engage with the change, ultimately coming to a degree of acceptance and adoption. Employees start being able to articulate "What's in it for me?", with the benefits of the change coming into realisation. They are likely to start experimenting, exploring the new ways of working, making adjustments based on feedback and experiences, gradually becoming more comfortable. For example, after an organisational restructure, employees experiment with how to get things done within the new reporting lines.

Some neurodivergent individuals thrive on making changes and experimenting in this way, as it feeds their creative thinking. They may naturally take the lead in trying new ways and taking risks. Others have more perfectionist tendencies, which makes experimenting more stressful. It may seem risky, and they might want to protect themselves by not engaging with the experiments until they know the outcomes. As a manager, make it OK to fail and to find other paths forwards. Actively inviting an individual's problem-solving can support the group's experimentation process. Others may be better at incremental change and might have a role later in the process.

Workplace changes often result in changes to unwritten facets of workplace life: power dynamics, the politics of an organisation, or "how we get things done". Some neurodivergent individuals will find it difficult to understand changes to unwritten rules. Support them by assigning a mentor or buddy who can help them make sense of things.

As a manager, you can support your team in creating new habits and routines: for example, by adjusting formal process documentation or having an open discussion about how people want to work in the changed workplace e.g. "How do we want to share updates as a team in this new structure?"

When employees start to accept the change, their energy level is likely to increase. For example, after a software upgrade, employees recognise its efficiency and the impact of new, streamlined processes. Reinforce this positive progress via feedback and celebration of improvements that encourage the new setup. Neuroinclusive training, coaching and mentoring can also help embed the changes.

In the final part of Stage 3, integration, employees have fully incorporated the change into their daily routines. The new ways of working have become the norm, and the organisation moves forwards more efficiently, with employees finding new ways to contribute and to be of value.

Table 8.1 distils much of what we have covered in this chapter into a simple checklist.

Table 8.1 Planning checklist for a neuroinclusive change process

Have you...	*Completed*
...talked to HR about how these planned changes might impact some groups of people more than others?	
...provided (a) information verbally? (b) an opportunity to ask questions? (c) a written document in plain English summarising key points?	
...explained how the changes will impact day-to-day tasks, who they will impact and when they will happen?	
...given a rationale for the changes - the how and the why?	
...provided opportunities for the employees to be consulted, offer their input and ask questions, both in group forums and 1:1?	
...made assistance available from mentors, HR and external support with regard to accommodations/ adaptations to prepare for the change?	

Case study: Desk Moves

Changes to seating plans can cause a lot of unnecessary distress. I've been through many of these over the years and, when I was in senior roles and responsible for deciding the new plans, I took the time to ask everyone what their preference would be, trying to match the layout to preferences as best as possible. But I never experienced this from other senior staff when I was in a more junior role and found myself sitting in the middle of open-plan offices, which is a nightmare for my sensory processing issues.

49-year-old female senior manager in professional services

What else can I do to support others through change?

When change is happening, people tend to think about things that are out of their control, catastrophising about outcomes and then ruminating about what will happen next. To address this, try an exercise which focuses on what they *can* do in a situation, what choices they have, thereby reducing the likelihood of overwhelm. It also works well as a visual and physical exercise to get thoughts out your head.

Figure 8.2 presents the Area of Influence, which is divided into two parts:

1. **The Area of Influence.** In this area are things that you have more control over. It typically includes your own thoughts or what you direct your time and energy on.
2. **The Area of Concern.** This is everything else that is going on over which you have no control or influence. It might include a redundancy process, for example. These are the things that often simply need to be accepted.

Area of Influence[13]: Exercise

This can be completed in person through a facilitated conversation. Alternatively, invite a team member to do this on their own and then

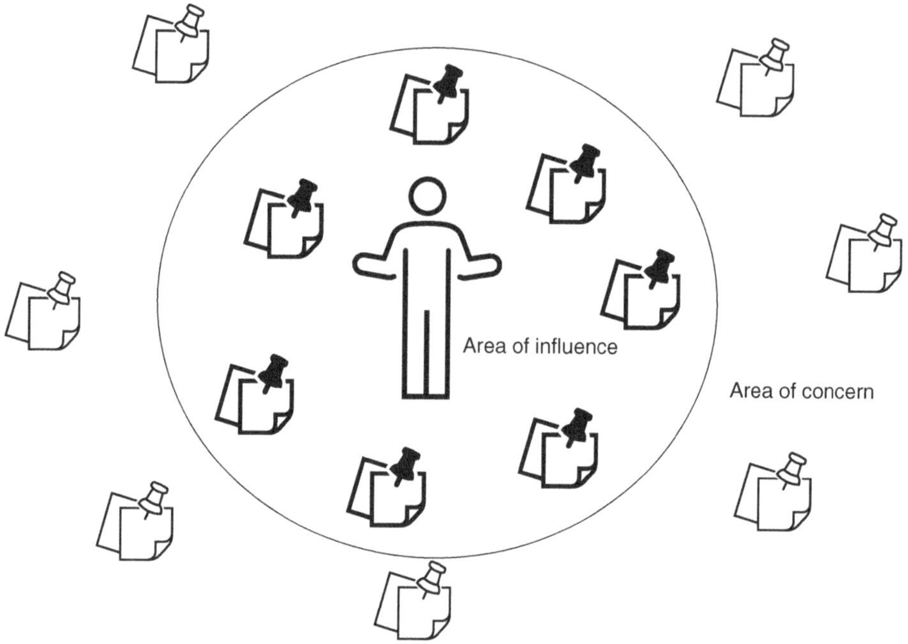

Figure 8.2 Area of Influence

share what they have discovered. You will need something to write on (A4 paper, flipchart or whiteboard), sticky notes and a pen (non-permanent for a whiteboard!).

1. Write everything you feel concerned about, in relation to this change, on sticky notes, one note for each thing. It can relate to any part of your life: work, home, friendships or personal life.
2. Stick all those sticky notes onto the paper or board.
3. Draw a smaller circle onto the area. Move into that circle those sticky notes that contain concerns over which you have some influence.
4. Reflect on the notes in the inner circle and one sticky note at a time. Ask yourself:

    a. What can I do in order to influence this?
    b. Who shall I speak to?
    c. What do I ask for?
    d. What do I believe about myself and them?
    e. How might I feel as a result of doing this?

People often think about the whole area of concern and feel overwhelmed and powerless. Identifying what things can be in the circle enables a more objective view of the situation and a feeling of being more in control.

It's your role as a manager to help employees differentiate between what is within their control and what isn't. A manager might also find ways to increase area of influence: for example, by matching them with someone like a mentor or sponsor to develop their influence, or by providing them with other information. This kind of support is particularly important for individuals with low self-esteem or confidence which can be a consequence of unsupported neurodivergence.

How do *you* manage change?

We'll finish this chapter with a self-reflection exercise that helps you evaluate your own capacity for managing change based on your personal characteristics and managerial style. Ask yourself how the answers align with the insights from this chapter.

What are your levels of optimism like when there is high uncertainty? Are you a "can do" person? Or do you tend to critique and look for problems?

....

....

....

....

Do you bounce back and reframe quickly? Or do you take a long time to adapt?

....

....

....

....

What happens for you when you believe others are leaning into victimhood or overreacting?

....

....

....

....

How do you prefer others to respond to change?

....

....

....

....

What is yours to do as a manager?

....

....

....

....

What can you do to model self-care and recovery time during a period of change?

....

....

....

....

Change provides an opportunity for teams to grow and learn: it can bring them together, strengthen relationships and develop connections. But this can happen only so long as they are adequately supported and are willing to share their vulnerabilities and concerns openly. As well as looking after your team through a period of change, you must also look after yourself.

Key points from Chapter 8

1. Change is a constant in today's workplaces, whether from external or internal sources. Technology has made the rate of change exponential, so equipping ourselves is crucial.
2. Companies are often well versed in managing the *process* of change but they may overlook the psychological and emotional impacts.
3. Attending to employees' emotional responses can smooth the transition to a post-change landscape, improve team dynamics and reduce conflict. It can also improve relations between team members and create a sense of belonging.
4. Neurodivergent individuals can experience change differently to other individuals. Recognise this by creating time and space for them.
5. Tools such as the Area of Influence and the EKR Change Curve® can help you support your team through change by enabling them to manage their emotions and influencing skills more effectively.

Notes

1 Graham, D.W. (ed.), "Heraclitus". *Stanford Encyclopaedia of Philosophy*, 2023. Available at: https://plato.stanford.edu/entries/heraclitus

2 Kelly, C., Martin, R., Taylor, R. and Doherty, M., "Recognising and responding to physical and mental health issues in neurodivergent girls and women". *British Journal of Hospital Medicine*, 85(4) (6 May 2024). https://doi.org/10.12968/hmed.2023.0337. Morie, K.P., Jackson, S., Zhai, Z.W., Potenza, M.N. and Dritschel, B., "Mood disorders in high-functioning autism: The importance of alexithymia and emotional regulation". *Journal of Autism and Developmental Disorders*, 49, 2935–45 (July 2019). https://doi.org/10.1007/s10803-019-04020-1. Kinnaird, E., Stewart, C.S., and Tchanturia, K., "Investigating alexithymia in autism: A systematic review and meta-analysis". *European Psychiatry*, 55 (January 2018), 80–89. https://doi.org/10.1016/j.eurpsy.2018.09.004

3 Equality and Human Rights Commission, "Assessing impact and the equality duty: An eight step guide". Equality and Human Rights Commission, 29 June 2020. Available at https://www.equalityhumanrights.com/guidance/public-sector-equality-duty/assessing-impact-and-public-sector-equality-duty-scotland-1

4 UK Government Equalities Office, "Public Sector Equality Duty: Guidance for public authorities". UK Government, 18 December 2023. Available at https://www.gov.uk/government/publications/public-sector-equality-duty-guidance-for-public-authorities/public-sector-equality-duty-guidance-for-public-authorities

5 Fragouli, E., "Leadership effectiveness for managing change: The role of emotions". *The Business and Management Review*, 10(1) (November 2018), 14-23. Available at: https://cberuk.com/cdn/conference_proceedings/2019-07-12-19-45-45-PM.pdf
6 Kübler-Ross, E., *On Death and Dying*. London: Tavistock Publications, 1970.
7 Roe, K., *Practice and Perspectives*. Oxford, UK: Oxford University Press, 2017.
8 Broadhurst, S., and Forrester-Jones, R., *Autism and Loss*. London: Jessica Kingsley, 2007.
9 Broadhurst, S., and Forrester-Jones, R., *Autism and Loss*. London: Jessica Kingsley, 2007.
10 Patros, C.H.G., Alderson, R.M., Tarle, L.J., Lea, S.E. and Hudec, K.L., "Choice-impulsivity in children and adolescents with attention-deficit/hyperactivity disorder (ADHD): A meta-analytic review". *Clinical Psychology Review*, 43 (2016), 162–74. https://doi.org/10.1016/j.cpr.2015.11.001
11 Leotti, L.A., Iyengar, S.S., and Ochsner, K.N., "Born to choose: The origins and value of the need for control". *Trends in Cognitive Science*, 14(10) (October 2010), 457–63. https://doi.org/10.1016/j.tics.2010.08.001
12 Luke, L., Clare, I.C.H., Ring, H., Redley, M. and Watson, P., "Decision-making difficulties experienced by adults with autism spectrum conditions". *Autism*, 16(6) (2011), 612–21. https://doi.org/10.1177/1362361311415876
13 Adapted from Covey, S.R., *The Seven Habits of Highly Effective People: Restoring the Character Ethic*. New York: Simon & Schuster, 1989, pp. 81–94.

9 Wellbeing

Everyone, depending on the point in their life, will find their headspace ranging from just about surviving, through to doing "OK", to thriving. Sometimes you feel positive, engaged and able to handle what life presents; other times you don't even have the motivation to get out of bed. Neurodivergent individuals are more likely to experience mental health conditions at the "surviving" end of this spectrum, such as anxiety, depression and burnout (see Figure 9.1), compared to neurotypical individuals.[1] While managers and leaders might not have direct control over all the factors (e.g. biological) driving these conditions, they can mitigate their severity and prevalence by helping to create a healthy and supportive workplace. In this chapter, we will discuss what actions managers can take to do this, with a specific focus on how to build your team's personal resources and manage demands.

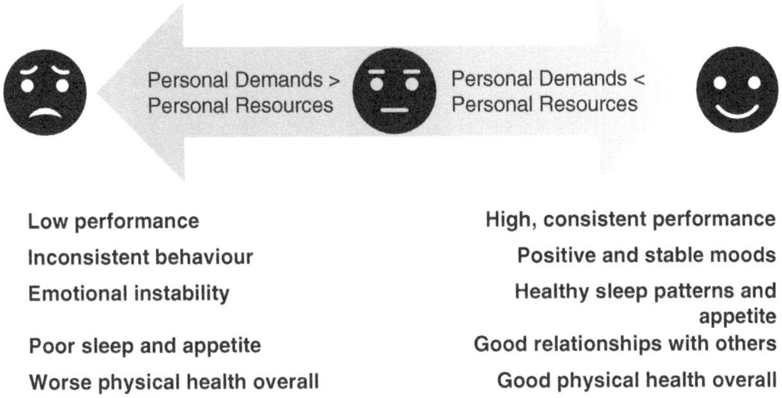

Figure 9.1 From surviving to thriving

DOI: 10.4324/9781003645535-9

Highs and lows of life

For this self-reflection exercise, look at Figure 9.2. Draw a line on the diagram that represents the course of your career – its highs and lows. It will fluctuate according to what kind of headspace you were in and what kind of energy or resources you had at each stage. It will range between good headspace and energy (thriving) and poor (surviving) (Y axis). Figure 9.3 is an example with the line already drawn on.

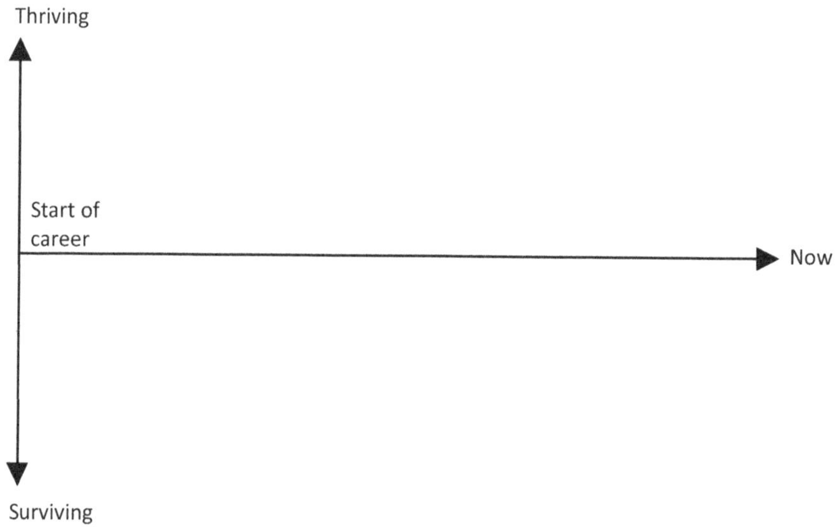

Figure 9.2 Highs and lows (blank chart)

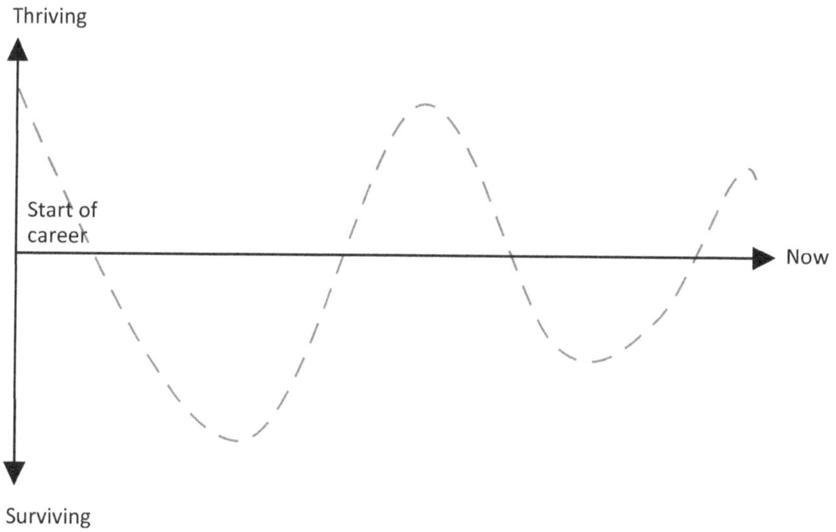

Figure 9.3 Highs and lows (example)

Look at the line you have drawn and identify four instances in your work career that were either peaks or troughs. It might include promotions, changes in management and personal circumstances such as illness in your family. In relation to these peaks or troughs, ask yourself the questions laid out in Table 9.1.

Table 9.1 Reflection on career peaks and troughs

Question	*In the times where you were thriving*	*In the times where you were only just surviving*
What support or resources did you have available to you personally, e.g. social circle, professional support?		
How well did the physical environment in the office work for you?		
How well did your manager's approach and support work for you?		
How did the way in which your job was defined support or hinder you?		
What personal demands or responsibilities did you have at the time e.g. family and finances?		
Which workplace systems (e.g. policies, procedures) worked for you and which didn't?		
How did you perceive your performance at work at the time?		

Perhaps this exercise has shown that there are both internal and external factors that contribute to you thriving, that make it more likely for you to be just surviving. These factors are different for everyone, but there are common themes. The following section covers more common themes that recur for neurodivergent individuals, and looks at how you can offer support as their manager.

Understanding wellbeing

"Wellbeing" has become an everyday - if not overused - word over the last ten years, so let's define what we mean by it in this book. We will use the World Health Organisation (WHO) definition, which defines it as "a positive state experienced by individuals and societies. Similar to health, it is a resource for daily life and is determined by social, economic and environmental conditions".[2]

In this chapter, we will see wellbeing as a balance at any one time between personal resources (which increase wellbeing) and personal demands (which reduce wellbeing), whereby having a good level of wellbeing is when personal resources are greater than personal demands. When demands significantly outweigh personal resources over a period of time, poor wellbeing can result, including mental health issues.

Mental health

In the UK, the Office for National Statistics[3] data on wellbeing indicates that it has remained static over time, when considering factors such as strength of relationships, employment, physical health and finances. However, one component of wellbeing has declined - mental health: the proportion of people experiencing depression or anxiety increased from 19.7% in 2017/18 to 23.7% in 2020/21.[4] Experts in this field have suggested that there has been an epidemic of mental health and burnout since the COVID-19 pandemic, particularly among certain populations.[5]

Generally, neurodivergent individuals are more likely to suffer mental health issues than neurotypical peers,[6] and estimates suggest that around 70% of neurodivergent individuals have to take time off work at some point for mental health reasons.[7] Let's look at ND populations more specifically:

- **Autism.** Studies[8] suggest that 6 in 10 autistics have considered suicide, 3 in 10 have made a suicide attempt and 8 in 10 have a mental health condition
- **ADHD.**[9] 30-40% of those with depression have ADHD (versus 10-15% of the general population having depression), and women

with ADHD are four times more likely to die by suicide than women in the general population
- **Dyslexia.** Those with dyslexia are more likely to have social anxiety than other populations[10]
- **Obsessive Compulsive Disorder.** 47.7% of individuals with OCD will also have a mood disorder, and 32% have anxiety[11]
- **Tourette's.** People with Tourette's have poorer mental health than those in the general population[12]
- **Dyspraxia.** A diagnosis of dyspraxia has been linked to higher rates of anxiety and depression[13]

It is thought that society's (negative) response to neurodivergent conditions and traits drive behaviours that impact mental health, such as masking (see Chapter 2), which has been linked to suicide, anxiety and depression.[14]

Additionally, neurodivergent conditions often have traits and co-occurring challenges that can also negatively impact mental health and wellbeing. For example:

- **Sleep disorders.** These are more common, which can impact mindset, energy and level of anxiety[15]
- **Disordered eating.** 35% of people with an eating disorder are thought to be autistic compared to around 1% in the overall population.[16] Furthermore, anorexia is more likely in autistic women[17]
- **Obesity.** Those with ADHD are more likely to be obese[18]
- **Under-employment.** As discussed in Chapter 1, employment contributes to wellbeing by providing human connection through workplace friendships, personal growth through skills development and training, and an opportunity to gain emotional support – plus the opportunity to be financially stable
- **Exhaustion.** Neurodivergent employees are more likely than their neurotypical peers to feel exhausted[19]
- **Loneliness.** Research shows that 23% of ND individuals are lonely at work, in contrast to 17% of their neurotypical peers[20]
- **Responses to stress.** ND individuals are more likely to feel excessive pressure; 35% report that they feel this way in contrast to a 29% level in neurotypical individuals[21]

- **Increased life-challenges.** Neurodivergent individuals are also more likely to experience difficulties such as divorce, trauma and addiction, which require additional support[22]
- **Perimenopause and menstrual health symptoms.** Associated with changes in hormones in women, symptoms include heightened sensitivities, depression, anxiety, emotional regulation and brain fog[23]

If managers and leaders are able to recognise the symptoms of poor mental health and wellbeing, they can help employees access the right resources at the right time. As a manager, you may already have access to training in this area, and HR policies in place. You may have mental health first-aiders or an employee assistance programme. Additional resources can be found in Chapter 12. Individuals may not want to disclose details of their mental health issue to their manager (see Chapter 3 on prejudice and biases), so signposting to a confidential neutral space could be more appropriate.

Personal resources

This section offers you techniques to help your employee tap into their own personal resources, which can increase their wellbeing and reduce the risk of poor mental health.

1. **Partner with them to develop coping strategies.** As we saw in Chapter 7, a focus on strengths can improve wellbeing and performance. Nonetheless, in the reality of working life most people will be faced with tasks they will find difficult. Many neurodivergent individuals develop coping strategies which they can apply in the workplace, such as time management techniques to minimise distractions and maximise the opportunity to hyperfocus.
2. **Identify their motivation and tap into it.** In *Drive*,[24] Dan Pink talks about three forms of intrinsic motivation:
   a. Autonomy. The desire to have control over one's work. Balancing their need for autonomy with your need to oversee their work. See Chapter 5 for more on this topic.

b. Mastery. The desire to excel at something meaningful. You can help develop this by providing stretch opportunities, to learn and develop skills that use their strengths (see Chapter 7). Alternatively, explore how to make the most of their skills within their role or the company.
c. Purpose. The need to contribute to something greater than oneself. Articulating projects and initiatives in the context of a broader life purpose can increase motivation. For example, if you invite an ADHD individual to set up an employee resource group, you could frame it in terms of the impact it might have on other people like them.

3. **Create opportunities to work flexibly.** Many neurodivergent individuals face challenges with sleep, some at a diagnosable level:[25] for instance, an estimated 80% of autistic individuals have sleep problems.[26] As most people know from experience, even a short-term lack of sleep can cause changes in behaviour and mindset, e.g. finding the workplace more demanding. For neurodivergent individuals, this can include reduced social skills, a negative impact on memory and a reduced ability to interpret social cues.[27] In the general population, long-term lack of sleep is linked to poor health and, for chronic sufferers, decreased life expectancy.[28]

   Managers who are aware of the potential challenges with sleep can remind team members of the offer and role-model flexible working. Ask what strategies they already have in place and, if asked for advice, you can offer suggestions such as setting and maintaining a sleep schedule, being outdoors and taking exercise.[29]

4. **Find ways to have wellbeing check-ins.** Agree a way to check in regularly - particularly if you are in different locations - to maintain an open and regular dialogue about how they are doing. This makes it easier to have discussions about what kind of support they might require. There are many techniques to initiate a wellbeing check-in; these two are among the most common:
   a. **The blob tree.** This allows people to express how they are feeling visually. It's helpful for those who don't like talking about feelings

or who are less emotionally literate. There are many versions of the blob tree but they all feature gender-neutral characters spaced around a tree in different guises, e.g. climbing the tree, sliding down it, helping each other, dancing, etc. Ask them what character they most identify with; you could begin by disclosing what character you identify with as a way of building trust

b. **Spoons.**[30] Energy management can be challenging for neurodivergent individuals; even talking about energy levels can be difficult. Exhaustion is more common for them.[31] Spoon theory asks us to imagine that each person has a finite number of spoons. Certain tasks deplete spoons and others replenish them. For example, an important meeting costs two spoons, while having some downtime restores them. So, when someone is low on energy, they could describe themselves as being "low on spoons", a shorthand that allows a discussion of energy levels but avoids going into detail and saves embarrassment

Techniques for employees to increase their energy levels include:[32]

- Shifting attention between energy-draining and energy-increasing tasks. A "strengths sandwich" is where an employee uses a strength on a task, follows it with a more challenging task, and then undertakes a task that uses another strength (see Chapter 7 for more on strengths)
- Creating achievable short- and medium-term goals to engender positive feelings of achievement
- Helping co-workers with their tasks – particularly good for those who are motivated by helping others
- Doing micro-meditations, e.g. closing your eyes for three minutes at your desk (see Chapter 12 for resources around this)
- Taking a walk outside, particularly in a green space
- Having a 20-minute power nap
- Having times of the day or week when there are no meetings or interruptions

Neurodivergent individuals who are parents and caregivers are also more prone to burnout,[33] so may benefit from additional support from

HR policies (such as carer's leave) and resources for working families (see Chapter 12).

Personal demands

Demands on the individual may well be self-imposed ones that derive from neurodivergent traits, behaviours or personality characteristics. Prevalent ones are:

Perfectionism. There are two kinds of perfectionism: "front-end" and "back-end". Front-end perfectionism prevents individuals from starting tasks until they feel equipped to do it all.[34] It has been suggested that this is a significant issue for many neurodivergent individuals. If you are managing a front-end perfectionist, discuss what support they might need to complete the task. Back-end perfectionism means tasks might not be completed in a timely way because of overly high standards. If you are managing such a perfectionist, help by setting clear expectations about what can be done to 80%, or 90%, perfection; only what you both agree is important should be done to 100% perfection. Make it OK to fail or make errors (see Chapter 7).

People pleasing. Boundaries may also be overstretched when there is an excessive need to please others over and above one's own wellbeing. This is common in those with rejection sensitivity dysphoria (see Chapter 2),[35] who may deploy this strategy to reduce the chances of feeling rejected.[36] Work with them to increase the trust that they have in themselves, and support the capabilities they have to break that behaviour. Coaching can help achieve this.

Imposter syndrome. Imposter syndrome is the persistent inability to believe that one's success is deserved or has been legitimately achieved through one's own efforts or skill. It is thought that neurodivergent individuals can experience high levels of self-criticism, which may contribute to higher levels of imposter syndrome in the ND population.[37] Techniques that can help include actively supporting their ideas and contributions, talking openly about imposter syndrome, sharing examples of people who have overcome imposter syndrome (being mentored by them) and challenging the individual's ideas about competence (which are sometimes are incorrect or unrealistic).[38]

Overthinking. You can help break a rumination cycle by encouraging an overthinker to do something that requires them to pay complete attention, e.g. standing on one leg and counting backwards.[39] Changing the environment can also help them to move on mentally, so take the time to go outside for a 1:1, for example.

Masking. Masking has appeared as a topic a number of times already (see above and Chapters 2, 5 and 6). The key is to create a trusting environment in which individuals are comfortable in being themselves. They will have ideas about what they want to see and feel in order to be more authentic.

These patterns of behaviours are not mutually exclusive; they may overlap and contribute to one another. They may also fluctuate with time – for example, increasing in prevalence and severity in a time of high pressure or change. They can also be prevented by improvements in wellbeing and mental health.[40]

Case study: How individuals can thrive

These case studies highlight the role of managerial relationships and support in enabling team members to thrive.

> I once worked myself to complete burnout and made myself quite ill. My manager, who was the CFO, told me to go home and not come back until the New Year, which was six weeks away. This was fully paid. He had recognised that the only way I would rest and recover was his insistence that I take the time off and not try to stay in contact with my team during that time. I immediately became even more unwell, as I felt I had permission to acknowledge the burnout rather than trying to fight my way through it. The six weeks of enforced rest meant I returned much stronger and healthier. He remains the best manager I ever worked for.
>
> 49-year-old female senior manager
> in professional services

> I have learned to mask and manage my personal issues by internalising my stress to get by, but it has had a significant impact on my mental health. I have also tried to learn to 'sit' with ambiguity; I sort of shut it out until I can gain more clarity. When confronted with multiple things that need my attention, I now try to prioritise them and focus on one at a time, rather than trying to multi-task.
>
> 58-year-old male AuDHD in the healthcare sector

Key points from Chapter 9

1. Neurodivergent individuals face more challenges than their neurotypical peers in sustaining wellbeing in day-to-day life.
2. While it is not your role as a manager to fix these challenges, you can help provide the environment and support that helps to enable individuals to thrive.
3. Improved wellbeing can help prevent some consequences and symptoms of neurodivergence: it supports the building of personal resources and management of demands.
4. These techniques should be supplemented with external professional help when required. This can be facilitated through HR.
5. At the core remains the relationship between manager and employee, based on trust, empathy and compassion.

Notes

1 Dow, D., Morgan, M., Hooker, J.L., Michaels, M.S., Joiner, T.E., Woods, J. and Wetherby, A.M., "Anxiety, depression, and the interpersonal theory of suicide in a community sample of adults with autism spectrum disorder". *Archives of Suicide Research*, 25(2) (April–June 2021), 297–314. https://doi.org/10.1080/13811118.2019.1678537. Wallis Towers Watson, "Almost three quarters of neurodiverse employees suffering with mental health issues". Wallis Towers Watson, Press release, 6 September 2024. Available at: https://www.wtwco.com/en-gb/news/2022/09/almost-three-quarters-of-neurodiverse-employees-suffering-with-mental-health-issues. Sharma, E., *et al.*, "Comorbities in obsessive-compulsive disorder across the lifespan: A systematic review and meta-analysis". *Frontiers of Psychiatry*, 12 (November 2021). https://doi.org/10.3389/fpsyt.2021.703701

2 World Health Organisation, "Health Promotion Glossary of Terms 2021". World Health Organisation, 2021. Available at: https://www.who.int/publications/i/item/9789240038349
3 Office for National Statistics, "UK Measures of National Well-being Dashboard". UK Government, 2025. Available at: https://www.ons.gov.uk/peoplepopulationandcommunity/wellbeing/articles/ukmeasuresofnationalwellbeing/dashboard
4 Office for National Statistics, "UK Measures of National Well-being Dashboard". UK Government, 2025. Available at: https://www.ons.gov.uk/peoplepopulationandcommunity/wellbeing/articles/ukmeasuresofnationalwellbeing/dashboard
5 Mental Health UK, "Burnout Report 2025 reveals generational divide in levels of stress and work absence". Mental Health UK, 16 January 2025. Available at: https://mentalhealth-uk.org/blog/burnout-report-2025-reveals-generational-divide-in-levels-of-stress-and-work-absence/. World Health Organisation, "COVID-19 pandemic triggers 25% increase in prevalence of anxiety and depression worldwide". World Health Organisation, 2 March 2022. Available at: https://www.who.int/news/item/02-03-2022-covid-19-pandemic-triggers-25-increase-in-prevalence-of-anxiety-and-depression-worldwide BBC News, "Rise in psychological distress in young adults: Survey". BBC, 6 July 2023. Available at: https://www.bbc.co.uk/news/health-66122386
6 Kirby, A., "Is there a link between neurodiversity and mental health?" *Psychology Today*, 26 August 2021. Available at: https://www.psychologytoday.com/gb/blog/pathways-of-progress/202108/is-there-a-link-between-neurodiversity-and-mental-health.
7 Wallis Towers Watson, "Almost three quarters of neurodiverse employees suffering with mental health issues". Wallis Towers Watson, Press release, 6 September 2024. Available at: https://www.wtwco.com/en-gb/news/2022/09/almost-three-quarters-of-neurodiverse-employees-suffering-with-mental-health-issues
8 Autistica, "Understanding suicide in autism". Autistics, n.d. Available at: https://www.autistica.org.uk/our-research/research-projects/understanding-suicide-in-autism. Hollocks, M.J., Lerh, J.W., Magiati, I., Meiser-Stedman, R. and Brugha, T.S., "Anxiety and depression in adults with autism spectrum disorder: A systematic review and meta-analysis". *Psychological Medicine*, 49(4) (2019), 559–72. https://doi.org/10.1017/S0033291718002283
9 Cumyn, L., French, L., and Hechtman, L., "ADHD and comordbid depression". *Current Attention Disorders Reports*, 1(2) (May 2009), 53–59. https://doi.org/10.1007/s12618-009-0008-1 Babinski, D.E., Neely, K.A., Ba, D.M. and Liu, G.L., "Depression and suicidal behavior in young adult men and women with ADHD: Evidence from claims data". *Journal of Clinical Psychiatry*, 81(6) (2020), 7–13. https://doi.org/10.4088/JCP.19m13130. Owens, E.B., Zalecki, C., Gillette, P., *et al*., "Girls with childhood ADHD as adults: Cross-domain outcomes by diagnostic persistence". *Journal of Consultation and Clinical Psychology*, 85(7) (2017), 723–36. https://doi.org/10.1037/ccp0000217. Dodson, W., "The ADHD–depression link: Symptom parallels and distinctions". *Additude*, 4 August 2025. Available at: https://www.additudemag.com/adhd-depression-link-symptoms-diagnosis-treatments/amp/

10 Georgiou, G.K., Parrila, R. and McArthur, G., "Dyslexia and mental health problems: Introduction to the special issue". *Annals of Dyslexia*, 74 (2024), 1–3. https://doi.org/10.1007/s11881-024-00300-3

11 Sharma, E., *et al.*, "Comorbidities in obsesssive-compulsive disorder across the lifespan: A systematic review and meta-analysis". *Frontiers of Psychiatry*, 12 (November 2021). https://doi.org/10.3389/fpsyt.2021.703701

12 Steft, M.E., "Mental Health needs associated with Tourette syndrome". *American Journal of Public Health*, 74(12) (1984), 1310–13. https://doi.org/10.2105/ajph.74.12.1310

13 Bidwell, V., *Developmental Coordination Disorder (Dyspraxia): What Helps, What Hinders in the School Years for Later Achievement and Wellbeing?* Doctoral thesis (D.Ed.Psy), UCL (University College London), 2022.

14 Mylett, M., *"I wish I could just be myself": Social Camouflaging, Internalized Stigma, and Internalizing Mental Health Problems in Adult ADHD*. MA thesis, Simon Fraser University, 2023. Available at https://summit.sfu.ca/_flysystem/fedora/2023-02/etd22306.pdf. Evans, J.A., Krumrei-Mancuso, E.R. and Rouse, S.V., "What you are hiding could be hurting you: Autistic masking in relation to mental health, interpersonal trauma, authenticity, and self-esteem". *Autism in Childhood*, 6(2) (June 2024), 229–40. https://doi.org/10.1089/aut.2022.0115. Miller, D., Rees, J. and Pearson, A., "'Masking is life': Experiences of masking in autistic and non-autistic adults". *Autism in Adulthood*, 3(4) (December 2021),a 330–38. https://doi.org/10.1089/aut.2020.0083. Hull, L., Levy, L., Lai, M.C., *et al.*, "Is social camouflaging associated with anxiety and depression in autistic adults?" *Molecular Autism* 12(1) (2021): 13. https://doi.org/10.1186/s13229-021-00421-1

15 Prince, T.-M., and Abel, T., "The impact of sleep loss on hippocampal function". *Learning and Memory*, 20(10) (October 2013), 558-69. https://doi.org/10.1101/lm.031674.113. Colten, H.R., and Altevogt B.M. (eds.), *Sleep Disorders and Sleep Deprivation: An Unmet Public Health Problem*. Institute of Medicine (US) Committee on Sleep Medicine and Research; Washington (DC): National Academies Press, 2006: ch. 4. Available at: https://www.ncbi.nlm.nih.gov/books/NBK19958/. Törnroos, M., Hakulinen, C., Hintsanen, M., Puttonen, S., Hintsa, T., Pulkki-Råback, L., *et al.*, "Reciprocal relationships between psychosocial work characteristics and sleep problems: A two-wave study". *Work & Stress*, 31(1) (2017), 63-81. https://doi.org/10.1080/02678373.2017.1297968

16 Autistica, "Treating anorexia in autistic women". Autistica, n.d. Available at: https://www.autistica.org.uk/our-research/research-projects/treating-anorexia-in-autistic-women

17 Autistica, "Treating anorexia in autistic women". Autistica, n.d. Available at: https://www.autistica.org.uk/our-research/research-projects/treating-anorexia-in-autistic-women

18 Cortese, S., *et al.*, "Association between ADHD and obesity: A systematic review and meta-analysis". *American Journal of Psychiatry*, 173(1) (August 2015). https://doi.org/10.1176/appi.ajp.2015.15020266

19 CIPD, *Neuroinclusion at Work Report 2024*. CIPD, 2024. Available at https://www.cipd.org/en/knowledge/reports/neuroinclusion-at-work

20 CIPD, *Neuroinclusion at Work Report 2024*. CIPD, 2024. Available at https://www.cipd.org/en/knowledge/reports/neuroinclusion-at-work

21 CIPD, *Neuroinclusion at Work Report 2024*. CIPD, 2024. Available at https://www.cipd.org/en/knowledge/reports/neuroinclusion-at-work
22 Dykshoorn, K.L., "Trauma-related obsessive-compulsive disorder: A review". *Health Psychology and Behavioral Medicine*, 2(1) (2014), 517–28. https://doi.org/10.1080/21642850.2014.905207. Ford, J.D. and Connor, D.F., "ADHD and posttraumatic stress disorder". *Current Attention Disorder Reports*, 1 (2009), 60-66. https://doi.org/10.1007/s12618-009-0009-0. Brown, M., Brown, S., Briggs, E.D., Germán, M., Belamarich, P.F. and Oyeku, S.O., "Associations between adverse childhood experiences and ADHD diagnosis and severity". *Academic Pediatrics*, 17(4), 349–55. https://doi.org/10.1016/j.acap.2016.08.013. Hoover, D.W., "The effects of psychological trauma on children with autism spectrum disorders: A research review". *Review Journal of Autism and Development Disorders*, 2 (2015), 287–99. https://doi.org/10.1007/s40489-015-0052-y. Hartley, S.L., Barker, E.T., Seltzer, M.M., Floyd, F., Greenberg, J., Orsmond, G. and Bolt, D., "The relative risk and timing of divorce in families of children with an autism spectrum disorder". *Journal of Family Psychology*, 24(4) (August 2010): 449–57. https://doi.org/10.1037/a0019847
23 Gottardello, D. and Steffan, B., "Fundamental intersectionality of menopause and neurodivergence experiences at work". *Maturitas*, 189(108107) (2024), 1–6. https://doi.org/10.1016/j.maturitas.2024.108107. Moseley, R.L., Druce, T., and Turner-Cobb, J.M., "Autism research is 'all about the blokes and the kids': Autistic women breaking the silence on menopause". *British Journal of Health Psychology*, 26(3) (September 2021), 709-26. https://doi.org/10.1111/bjhp.12477. Osianlis, E., Thomas, E.H.X., Jenkins, L.M. and Gurvich, C., "ADHD and sex hormones in females: A systematic review. *Journal of Attention Disorders*, 0(0) (2025). https://doi.org/10.1177/10870547251332319
24 Pink, D.H., *Drive: The Surprising Truth About What Motivates Us*. New York: Riverhead Books, 2009.
25 Al Lihabi, A., "A literature review of sleep problems and neurodevelopment disorders". *Front. Psychiatry*, 14:1122344. doi: 10.3389/fpsyt.2023.1122344
26 Al Lihabi, A., "A literature review of sleep problems and neurodevelopment disorders". *Front. Psychiatry*, 14:1122344. doi: 10.3389/fpsyt.2023.1122344
27 Al Lihabi, A., "A literature review of sleep problems and neurodevelopment disorders". *Front. Psychiatry*, 14:1122344. doi: 10.3389/fpsyt.2023.1122344
28 Medic, G., Wille, M. and Hemels, M.E., "Short- and long-term health consequences of sleep disruption". *Nature of Science and Sleep*, 19(9) (May 2017), 151-61. https://doi.org/10.2147/NSS.S134864
29 Becker, S., "Sleep problems in teens with ADHD: Causes and solutions", *Additude*, 11 May 2025. Available at: https://www.additudemag.com/sleeping-problems-teens-adhd-causes-solutions/?srsltid=AfmBOoqAd_EpOd-KpwWOpDy8cEZ-FhqNyTUEQpNOC7gQDHkmshcUVRYB. Shanahan, P.J., Isaac, M. and Blackwell, J.E., "Sleep disorders in attention-deficit hyperactivity disorder and autism spectrum disorder: A pragmatic approach to assessment and management". *BJPsych Advances*, 27(5) (2021), 320–32. https://doi.org/10.1192/bja.2020.65

30 Prange-Morgan, C., "What is 'Spoon Theory'? And why is it important?" *Psychology Today*, 17 April 2024. Available at: https://www.psychologytoday.com/us/blog/full-catastrophe-parenting/202403/what-is-spoon-theory-and-why-is-it-important
31 Tomczak, M.T. and Kulikowski, K., "Toward an understanding of occupational burnout among employees with autism: The Job Demands-Resources theory perspective". *Current Psychology*, 43 (25 February 2023), 1582–94. https://doi.org/10.1007/s12144-023-04428-0
32 De Bloom, J., Kinnunen, U. and Korpela, K., "Recovery processes during and after work: Associations with health, work engagement, and job performance". *Journal of Occupational and Environmental Medicine*, 57(7) (July 2015), 732–42. https://doi.org/ 10.1097/JOM.0000000000000475
33 Mental Health UK, *The Burnout Report*. Mental Health UK, January 2025. Available at: https://euc7zxtct58.exactdn.com/wp-content/uploads/2025/01/16142505/Mental-Health-UK_The-Burnout-Report-2025.pdf
34 Brown, E. and Morley, E., "Feeling like a fraud: Exploring how the impostor phenomenon manifests in neurotypical and neurodivergent first-year undergraduate students in the UK using the impostor phenomenon assessment". *Journal of College Student Mental Health*, 2025, 1–23. https://doi.org/10.1080/28367138.2025.2493896
35 Dodson, V., Modestino, E.J., Titiz Ceritoğlu, H. and Zayed, B., "Rejection Sensitivity Dysphoria in Attention Deficit/Hyperactivity Disorder: A case series". *Acta Scientific Neurology*, 7(8) (2024), 23–30. https://doi.org/10.31080/ASNE.2024.07.0762
36 Wilson, R.M., *The ADHDer's Guide to Saying No (Without Guilt): An ADHD-Friendly Way to Say No, Reduce Overwhelm, Manage Impulsivity, and Prioritize Your Well-being*. Self-published, 2025.
37 Brown, E. and Morley, E., "Feeling like a fraud: Exploring how the impostor phenomenon manifests in neurotypical and neurodivergent first-year undergraduate students in the UK using the impostor phenomenon assessment". *Journal of College Student Mental Health*, 2025, 1–23. https://doi.org/10.1080/28367138.2025.2493896
38 Hernandez, M and Lacerenza, C., "How to help high achievers overcome imposter syndrome", MIT Sloan Management Review, January 23, 2023, Available at: https://sloanreview.mit.edu/article/how-to-help-high-achievers-overcome-imposter-syndrome/.
Bates et al, "Overcome imposter syndrome: Contribute to working groups and build strong networks", *Biological Conservation*, Vol. 293, May 2024.
39 Hallowell, E., and Ratey, J., *ADHD 2.0: New Science and Essential Strategies for Thriving with Distraction*. London: Sheldon Press, 2025.
40 Müller, V., Mellor, D., and Pikó, B.F., "Associations between ADHD symptoms and rejection sensitivity in college students: Exploring a path model with indicators of mental well-being". *Learning Disabilities Research & Practice*, 39(4) (2024), 223–36. https://doi.org/10.1177/09388982241271511

10 Handling tricky situations

Sometimes the relationship breaks down between manager and employee, or employee and organisation. This chapter presents fictional case-study examples of crisis situations and suggests how they could be tackled, using the techniques and approaches covered in this book.

Every organisation, individual and environment is different. Therefore, there is no instruction manual on how to handle each situation. Rather, these examples are intended to get you thinking about tools and techniques to put in place to prevent these scenarios from materialising in your own workplace.

Relationship breakdowns

Let's begin with another self-reflection exercise.

What situations have you witnessed at work or experienced yourself, where a relationship has broken down between an employee and their manager?

....

....

....

....

What were the outcomes of those situations? For the manager? The employee? The organisation?

....

....

DOI: 10.4324/9781003645535-10

....

....

How would you articulate the root cause of those situations? For example, what issues weren't dealt with in an adequate or timely way?

....

....

....

....

What environmental factors contributed to the situation? For example, what policies, procedures, cultural or external factors had an influence?

....

....

....

....

We now turn to our fictional case studies, each of which features you as a key player, who can take action to address the problem. In each of the scenarios that follow, think about the influence on the situation and its outcomes brought to bear by (a) the manager's behaviour, (b) the employee's behaviour and (c) the workplace environment.

Example 1: Mark

You are starting in a new role in a marketing company. The departing manager, Pete, your predecessor, has told you the following about one of your new reports, Mark.

Mark is a 32-year-old marketing executive and has disclosed his ADHD to his manager. His reputation is that he is energetic, full of innovative ideas and great at connecting with team members and stakeholders. His team members loved working with him in the past, and so often forgave him for some of the things he did that occasionally caused irritation, such as delivering on tasks late (assumed to be due to poor time management), being a bit disorganised, droning on in meetings and losing focus during long meetings. His team thought of him as an enthusiastic co-worker who produced quality work. However, over time, his line manager - i.e. Pete - grew increasingly frustrated with the missed deadlines and the disorganisation. Pete expressed this frustration through jokes and jibes in group settings.

For example, you hear that Pete used to interrupt Mark when he was speaking, saying things like "We don't have time for one of your tangents today" or "Can you actually focus for once?" He apparently told Mark "You need a support assistant" when Mark misplaced his notes in a meeting. The team laughed along, even though a couple of them later mentioned that they had felt awkward. While no one asked Mark about how he received the comments, one co-worker asked Pete after a meeting about the comments he had made about Mark; Pete's response was that he was just encouraging Mark to do what was expected of someone at his level: "It's what he's paid for".

You know from your handover discussions that Pete felt that he couldn't trust Mark to deliver on time, as Mark had failed to meet a number of deadlines which had consequences for Pete. He responded by giving Mark shorter deadlines so he would have a buffer if he didn't deliver. Knowing some of the detail, you have a sense that these deadlines were too short - unrealistic - and therefore set Mark up for further failure. Other team members have told you that the tension between the two increased over time; Pete started leaving Mark out of projects, justifying it to Mark, when he asked why he hadn't been included, with "We need people who can multitask and get things done quickly, and not get distracted".

You learn that Mark appeared to lose self-confidence over time, unable to change the dynamic and focus on the tasks at hand because of the energy he spent worrying about the situation. The team noticed that he'd been quieter in the last few months, hesitating to contribute ideas and no longer engaging in casual conversations with colleagues. He mentioned to a couple of team members that he's been having frequent headaches and trouble sleeping, and that he is contemplating going on stress leave.

Pete has recently moved to head up another department and, as Pete's replacement, you are now Mark's manager. You have met with several of the team members and with Pete. You are about to go into your first meeting with Mark.

What do you see as the causes of this situation?

....

....

....

....

What other contributing factors are there?

....

....

....

....

How could the team have supported Mark more effectively?

....

....

....

....

How might you approach Mark as his new manager?

....

....

....

....

Three points to take away

1. So far, you have only heard information second-hand. Your first step is to listen to Mark and understand his perspective. You could help build trust, and demonstrate authenticity and empathy, by sharing a situation or two in which you were under pressure yourself, and what happened for you. Ask what he needs from you to be able to do his job well, including mental health support and reasonable accommodations. (See Chapters 4 and 5 for more on this.)
2. There are indications of low psychological safety in this team - people not speaking up. You have an opportunity to reset the tone of the team - its culture. Start by having a team discussion about everyone's expectations of one another (i.e. what behaviours are important), including how they might deal with unwanted behaviours when they see them. A team coach or facilitator can help here.
3. Demonstrate your commitment to and support for Mark's career by providing skills training, mentoring or coaching. It may also increase his confidence and create a healthy dynamic between the two of you.

Example 2: Emma

Emma is a senior project manager at a mid-sized technology company. You are her manager.

Known for her efficiency and results-driven mindset, she values direct communication and has little patience for what she considers unnecessary pleasantries. Her approach helps streamline decision-making, and some co-workers appreciate her directness and efficiency. However, it also creates friction within her team, where many prefer a more collaborative and relationship-oriented style of interaction. You don't particularly mind the direct approach, and find her honesty refreshing. She hasn't disclosed any neurodivergent condition to you.

New team members have recently joined and they aren't used to Emma's communication style. It has become a source of tension in a number of situations: the abrupt way she starts meetings; the directness of her feedback; her undiplomatic and factual contributions in discussions.

Emma prefers to dive straight into business, skipping small talk entirely. In one team meeting, she opened with "Let's get started. We're behind on the product launch timeline, and I need answers on where the delays are happening". Efficient though it might be, her team members found it rude and perceived it as setting a negative tone. It didn't allow them to settle into the meeting by talking about their plans for the weekend or recent football matches. Her abrupt approach also inhibited some team members from approaching her, believing that she would "bark" at them. Over time, this lack of casual engagement made team members less comfortable speaking openly in meetings.

As a manager, Emma is expected to provide feedback to her team. She often delivers it in a very matter-of-fact way, which gets the key points across but is perceived as being unempathetic. In one performance review, she told a colleague "Your reports are too slow, and they're not detailed enough. You need to fix this". Her colleague felt disheartened by the comment: he had been putting in extra hours to ensure accuracy, but Emma's phrasing made him feel like a failure rather than someone who needed guidance.

Emma also has a tendency towards perfectionism, and encourages others to meet those same standards. When they don't, she thinks it best to tell them straight. In one meeting, a colleague suggested an idea and she responded with "That won't work", with no further context. The colleague felt attacked and undermined.

You have tried providing feedback to her about this, but it doesn't appear to be working.

Emma's direct communication style has resulted in friction between her and the team. Some have even started avoiding her when possible, and the perceived lack of support from her has led to decreased morale. Several key members have started asking around for roles in other teams.

What are the root causes of this situation?

....

....

....

....

What other contributing factors are there to the team members wanting to leave?

....

....

....

....

What can be done by you as her manager at this point to resolve the situation?

....

....

....

....

What can you learn from this situation to apply to future situations in your workplace?

....

....

....

....

Three points to take away

1. Start by understanding Emma's perspective and try the Action-Impact-Do Differently (AID) feedback model to structure a conversation about what behaviours you would like her to change, and how she might do that. (See Chapter 5 for more on the AID model.) Assign her a mentor as an impartial sounding board as she seeks to shift her approach.
2. Emma has not disclosed any neurodivergent condition, so it is not possible to have a conversation with her about neurodiversity, even if she demonstrates traits. Feedback therefore needs to focus on the behaviour that she is demonstrating.
3. While you work with Emma to soften her tone and approach, role-model the behaviours that you want Emma to demonstrate and share what you appreciate about the wider team members.

Example 3: Storm

Storm (they/them), a 28-year-old autistic employee, works as a team leader in a client service team. They are highly detail-oriented, great at process optimisation and a tenacious problem-solver. Their manager, Sarah, a fast-paced and intuitive thinker with very little time, has absolute confidence in Storm's ability to get the tasks done and leaves them to do their job. She believes that Storm is a high-potential individual, and she has been working with Storm to navigate a potential move to another team so that they can develop their skills. You are Sarah's manager and Storm has been open with you and others about their autism.

In order to identify a new team, Storm has been told to network with more senior leaders. Storm initially tried going to networking events but found them overwhelming; making small talk is too difficult for them. So Storm is now relying on one-to-one interactions to build their network. They have set up a one-to-one meeting with you to introduce themself and ask you for support in looking for a role within your broader team. However, when Storm dialled into the call they switched their camera off, leaving you wondering what Storm was doing – were they multitasking or not interested in building a relationship with you? You also felt that Storm didn't share anything personal about themself, so you left feeling none the wiser about the kind of person they are, although you knew more about Storm's job achievements thanks to the long list they provided. As a result of this, you weren't inclined to recommend Storm for jobs.

Having met with you, Storm wanted to meet with a different senior manager, Bob. It was difficult to get a scheduled time slot with Bob, so instead Storm walked up to Bob when he was coming down the corridor, handed him a one-page summary of their achievements and said they wanted to meet soon. Bob was in the process of getting from one meeting to another, so was not in the headspace for Storm's enthusiastic intervention. The result was that Bob hurried away, making a mental note to try to avoid Storm in future.

Feeling stuck, Storm turned to Sarah and shared that they were desperate to move teams. Sarah told Storm to "Give me six months, and I'll work something out". Six months to the day, Storm asks Sarah what job she has arranged for them. Sarah is taken aback because she hadn't expected Storm to take her statement literally. Storm, having expected to move on within six months, resigns. Sarah has called you for advice, keen to prevent Storm from leaving.

What is the root cause of this situation?

....

....

....

....

What other contributing factors are there?

....

....

....

....

What could have been done to prevent this situation from occurring?

....

....

....

....

What advice would you give Sarah?

....

....

....

....

Three points to take away

1. Storm is struggling to identify and work within the social norms of the office. You can support them by ensuring that they have in place an individual or support network that can help them understand these unwritten rules: for example, an employee resource group or mentor (see Chapter 4) can help to raise their self-awareness.
2. Encourage Storm's manager to co-create a plan with Storm for how they can use their strengths to build their network with more senior managers.
3. Generate greater awareness of neurodiversity through training. Support this by acting as an inclusive leader and challenging bias as and when you see it (see Chapters 3 and 6).

Example 4: Jamie

Jamie is a talented graphic designer with a sharp eye for detail, working in the creative department of a mid-sized marketing agency. Jamie feels that his OCD and Tourette's diagnoses are a double-edged sword, bringing him his strengths in creativity and attention to detail but also making his day-to-day life exhausting as he tries to mask and manage his neurodivergent traits. You are his peer and work in the same team as him.

Jamie's OCD involves intrusive thoughts that lead to compulsive behaviours, such as rechecking designs or excessively reviewing emails to ensure perfection. This results in delays on deadlines, despite the high quality of his work. His manager – who is rarely visible – doesn't seem to notice the missed deadlines. Meanwhile, Jamie's involuntary grunting and repeating of words is distracting for some in the open-plan office, and – not wanting to raise it – a couple of people have started to work from home or have moved desks away from Jamie so that they can focus.

You love Jamie's work and have told him that you appreciate his contributions. However, team deadlines have been missed a lot recently as the workload for a particular project has increased, which has become very frustrating. While you have always intended to stay professional, sometimes your frustration has surfaced in curt feedback, leaving Jamie feeling hurt and lonely.

The dynamics within the team have also begun to shift. Initially, the team were understanding, often covering for delays and encouraging Jamie. Over time, however, tension has grown as others have covered Jamie's work as Jamie has fallen behind. Some colleagues subtly distanced themselves, while others made insensitive comments and jokes about Jamie's tics during informal conversations. Jamie, overwhelmed by a combination of his own struggles and the changing mood in the team, feels increasingly isolated. You have noticed a change in Jamie's demeanour and erratic behaviour; his tics have got worse. So you have set up a coffee date with him to talk.

As a peer, how might you approach your conversation with Jamie?

....

....

....

....

What is the root cause of the challenge in this case?

....

....

....

....

What other contributing factors are there?

....

....

....

....

Three points to take away

1. The key thing that might resolve this situation is open communication. Start by being curious about Jamie's perspective, offering him support while also providing a perspective on how his behaviours are impacting you as a colleague. Key to that conversation is a demonstration of empathy (see Chapter 5). Ensure that he has the right support structure from managers and peers to be able to perform at his best.
2. Encourage Jamie to talk to his manager about what is going on. Have a conversation about performance management. They may be able to provide reasonable accommodations in the office and for him to work from home. When talking to his manager, you could suggest that neurodiversity training is introduced to help increase the level of understanding in the team.
3. Work with Jamie to identify what other external support is available through occupational health, employee assistance programmes or employee resource groups (see Chapters 4 and 9).

Example 5: Ruth

You work in a high-profile professional services firm. You have just received a complaint from a key client regarding the quality of emails sent by one of the firm's remote team members, Ruth, who owns the client relationship. The client, an influential partner with a significant impact on future contracts, expressed concerns about multiple spelling mistakes, unclear phrasing and inconsistent communication in Ruth's emails and text messages. The client says that not only is it unprofessional, it causes confusion about the message being given. They believe that having someone "like Ruth" owning the relationship with them is a sign that you don't value their business. Given the importance of the client relationship, you need to find a way to calm the client down and prevent further irritation.

Ruth has disclosed dyslexia, which affects the accuracy and structure of her written communication. She is an expert in her field, and is great at building and maintaining relationships in person. However, Ruth now mainly works remotely so usually communicates with the team in writing. The remote work arrangement has already caused an issue with other team members, who perceive that she has been given "special treatment", as they are required to come into the office most days.

When the client lodged their complaint, they forwarded the emails in question. You can see what they mean: they are littered with spelling and grammatical mistakes and they fail to get the point across. They certainly don't look professional, and they aren't of the high standard that the company expects of employees.

You need to address this with Ruth, and with the client.

How might you approach this?

....

....

....

....

What is the root cause of the challenge in this case?

....

....

....

....

What other contributing factors are there?

....

....

....

....

Three points to take away

1. Set up a conversation with Ruth to share the client's feedback and to identify whether this is a one-off. Take responsibility for your part, as necessary. For example, did you notice mistakes and yet not provide feedback to Ruth? Once you have understood the client's perspective, feed back to Ruth and work together to improve the standard of written communication that she produces. Options include speaking to human resources about assistive technology, or the availability of an admin assistant to proofread emails before they are sent.
2. Having discussed the feedback with Ruth, work with her and the client to manage the relationship going forwards. Options could include changing the client manager or finding an outcome in which the client's standards can be met through accommodations.
3. Ruth may be open to other roles in which she can play to her strength of working with people in person and without as much written communication. Establish mentoring to look at such roles if Ruth is open to this. Alternatively, could the role be redesigned to reduce reliance on written communication, e.g. by pairing Ruth with someone else in a way that plays to both of their strengths (Chapter 7).

Example 6: Richa

Richa, a customer service agent at an insurance firm, has always been recognised for her enthusiasm and deep knowledge of policies. However, she has also struggled with managing her emotions, a trait of her diagnosed and disclosed ADHD. You are her manager.

While her passion for her work is evident, her reactions to stressful situations have often created disruption within the team. On multiple occasions, Richa has reacted strongly when colleagues disagreed with her approach to handling customer queries, which involved raising her voice or crying; this made other team members uncomfortable and sometimes upset. Additionally, her impulsivity leads her to sometimes interrupt others and talk over them which results in colleagues either not bothering to speak or else getting into arguments with her when she doesn't listen.

The situation has now reached a critical point following Richa's emotional outburst during a call with a high-value client. Frustrated by the client's repeated queries and perceived lack of understanding of the policies they were discussing, she responded abruptly by raising her voice. The client, unsettled by the interaction, immediately moved their business elsewhere. The loss of such a significant client has sparked concerns across the senior management team, raising urgent questions about how Richa's behaviour could be managed moving forward.

You haven't seen most of these situations first-hand, but rather heard the stories through others. That has deterred you from intervening or providing feedback to date, although now you appreciate that you have to do something.

Richa is aware that she has made a significant mistake, and didn't come into work the next day. You are concerned about her mental health as you know that she has struggled with depression in the past.

You are going to ask to meet Richa when she's back to work and you have been asked by the senior management team to sort the problem out.

What is the root cause of the challenge in this case?

....

....

....

....

What are the other contributing factors?

....

....

....

....

How might you approach the conversation with Richa?

....

....

....

....

Three points to take away

1. Given she is off work, you will need to manage the conversation with her sensitively. Talk to human resources or an expert in neurodiversity before you speak with her to provide information and advice in handling the situation. Identify what support might be available for Richa, and be prepared to offer it to her when you speak.
2. Having done this, you can increase understanding and regain confidence within the immediate and senior management team by providing information. Before you do this, agree with Richa what part of the plan you can share with senior management.
3. Over the coming months you could talk to the team about changes to the way they do their job, e.g. time away from phone enquiries, a few hours a day with guaranteed no interruptions, so that everyone benefits from such changes.

Key points from Chapter 10

1. Communication is key. Maintaining an open, empathetic and supportive dialogue, in a timely manner, is the foundation of good management.

2. Give people the space to share what is going on for them and to express their opinion. That way fewer assumptions are made, and things left unsaid are less likely to erupt into a crisis.
3. Use the community and support systems around you. Employee resource groups, human resources teams and occupational health can all provide resources, support and guidance which can prevent tricky situations from escalating.
4. Be the first to own what you now know and what you could have done differently. Perhaps it wouldn't have changed the outcome, but learning and not taking guilt into your next encounter is important. Don't beat yourself up if there isn't a simple solution: it often takes time and repeated attempts to work through complex people issues.
5. Managing performance and being clear about outcomes is key and should not prevent open conversations with neurodivergent team members.

11 In conclusion

Over millennia, humans have connected with and learnt from each other through stories. Stories – of love, fear, desire, disaster and many other things – are what shape our reality and tell us what is acceptable and what is not. Throughout this book, we have shared the stories and perspectives of many neurodivergent individuals whose voices provide an often unheard narrative – of a people who are not served well by our 21st-century ways of working. With this book we have brought information, reflective exercises, resources and perspectives for managers and leaders so that those stories can be heard, and so that our workplaces are adjusted to make sure the stories are different in the future.

In fact, the starting point is where managers *listen to* the stories of their neurodivergent team members – with empathy, compassion and curiosity. Acquiring an empathetic and compassionate approach to people management can be made easier by learning about neurodiversity: what it is, how it presents and the kinds of challenges that your neurodivergent team members face. This is what you have started to do by reading this book. It offers you a different lens: one through which they experience the world.

Creating an environment in which everyone can thrive starts with you: your behaviours, words, actions and how you role-model what inclusion looks like. It's your job to make everyone feel safe enough to bring their full selves to work. Openness, vulnerability and trustworthiness need to be at the core of all of your interactions. Challenge

DOI: 10.4324/9781003645535-11

your biases and assumptions, through self-reflection and feedback from others: there have been lots of self-reflection exercises in this book which will hopefully have helped you.

But bias is also reflected in the systems and processes around us, much of which is inherited from historic, less inclusive cultures. You may not be able to change these immediately, but there are always steps you can take. Start with what you can change – how you conduct interviews, for example.

This is not to say that creating an inclusive workplace is easy. Managers already report that their plates are full of things they need to change,[1] and addressing your own behaviour – which requires you to slow down and think about how you say and do things – adds more to the list. We hope that, by reading this book, you feel more confident and able to start creating that more inclusive environment, step by step. Continue to take steps forwards and, over time, your new practices will become your default way of operating.

To make these changes effective and sustainable, there are a few things to bear in mind. Firstly, have a clear understanding about your motivation and vision. We invited you to start thinking about this in Chapter 1. Now you've read the whole book, reflect on the questions below to help you achieve greater clarity about your motivations and what success might look like for you.

What kind of person would you like to be in your relationships with all of the people you work with?

....

....

....

....

What would that look like in terms of day-to-day behaviours?

....

....

....

....

What kind of team do you want to lead or manage?

....

....

....

....

How would that differ to the way you currently manage your team?

....

....

....

....

Having defined your motivations, the next step is to define actions that you can commit to right now. These don't need to be big. In fact, they are more likely to happen if they are small, everyday things. There is a list of suggested actions below; once you've done them, we encourage you to reflect on them and get feedback on how they've been received.

1. Share with your team, your peers and your manager the key take-aways for you from this book.
2. Ask your neurodivergent team members, or those you perceive of as thinking very differently to you, to share the things that they appreciate the most.
3. The next time there is a team discussion, instead of sharing what you know, focus on others' perspectives and views. Reflect on what you didn't expect and what you have learnt.
4. Ask each of your team members: "What one thing can I stop or start doing more of, to help you be at your best?"
5. Ask "What might I not be aware of which you would like to share with me to help your work?"

Choose just one of the above five actions - or another of your own - and commit to a time to complete it by. Write it here:

....

....

....

....

You may also want to consider how you can start to drive change within your broader organisation, by role modelling, championing change and pushing for systemic change.

Lastly, we thank you for taking the time and the effort to engage with our book and to reflect on your experiences. It is only by being aware of our behaviours and shifting our perspectives that we create workplaces in which everyone feels valued and supported, so that everyone can be more productive, more resilient and happier wherever we are.

Note

1 HR Daily, "Managers' plates 'piled higher and higher and higher'". *HR Daily*, n.d. Available at: https://www.hrdaily.com.au/news/11428

12 Resources

Neurodiversity-specific information and support

Description	*Link/details*
ADHD Foundation. An organisation working to improve awareness of ADHD and support ADHD individuals	www.adhdfoundation.org.uk
ADHD Adult UK. Resources including apps and peer groups	https://www.adhdadult.uk/
***Additude* magazine (US).** Online magazine with articles around ADHD	https://www.additudemag.com/
British Dyslexia Association. Information and support around dyslexia	https://www.bdadyslexia.org.uk/dyslexia
Dyspraxia Foundation. Organisation working to raise awareness of dyspraxia	https://dyspraxiafoundation.org.uk
The Codpast. Online publication for students and adults with dyslexia	https://thecodpast.org
Tourette's Action. Support and information about Tourette's	https://www.tourettes-action.org.uk/
The Dyscalculia Association. Promoting understanding and awareness of dyscalculia	http://www.dyscalculiaassociation.uk/
International OCD Foundation. Organisation raising awareness and support for those with OCD	https://iocdf.org
Embrace Autism. Autism screening including masking and empathy	https://embrace-autism.com/
National Autistic Society/Autistica. Information and support around the topic of autism	www.autism.org.uk

(*Continued*)

DOI: 10.4324/9781003645535-12

Description	*Link/details*
Indistractable. A book that helps people to be more focused by providing a step-by-step guide to reducing distractions	Eyal, N., *Indistractable: How to Control Your Attention and Choose Your Life.* London: Bloomsbury, 2019
ADHD Unpacked. A book that talks to the impact of ADHD diagnosis across an individual's life	Conner, A., and Brown, J., *ADHD Unpacked: Everything You Need to Know to Survive and Thrive as an Adult with ADHD.* London: Bloomsbury, 2025
Odd Girl Out. A book that talks to the experiences of a late-diagnosed woman with autism	James, L., *Odd Girl Out: An Autistic Woman in a Neurotypical World.* London: Bluebird, 2018
Speech and Language UK. Information and resources around developmental language disorder	https://speechandlanguage.org.uk/educators-and-professionals/dld-educational-support/
Neurodiversity for Dummies. Information and exercises around neurodiversity generally	https://www.dummies.com/book/body-mind-spirit/emotional-health-psychology/psychology/general-psychology/neurodiversity-for-dummies-301637/
ACAS. Information and advice around neurodiversity at work	https://www.acas.org.uk/neurodiversity-at-work
Example of online resources provided by a union including Public Sector Equality Duty	https://prospect.org.uk/topic/neurodiversity/
Examples of reasonable adjustments	https://www.ucu.org.uk/media/6091/Reasonable-adjustments---removing-barriers-to-disabled-people-at-work-UCU-guidance/pdf/Reasonable_adjustments.pdf

Employer/employee reference information

Area	*Description*	*Link/details*
Flexible working	UK government. Information around flexible working applications	https://www.gov.uk/flexible-working/applying-for-flexible-working

(Continued)

Area	*Description*	*Link/details*
Communication	Scottish government. Inclusive communication principles	https://www.gov.scot/publications/principles-inclusive-communication-information-self-assessment-tool-public-authorities/pages/5/
	UK government. Information around making written text more accessible	https://gcs.civilservice.gov.uk/guidance/accessible-communications/
Mental health resources	The Samaritans. For immediate emergency mental health support (alongside the emergency services, as relevant)	https://www.samaritans.org
	Mental Health First Aiders England. To train individuals on how to respond to mental health conditions in the workplace	https://mhfaengland.org/
	Mind (UK). Support for mental health conditions	https://www.mind.org.uk/
	Campaign Against Living Miserably (CALM). App for meditation and emotions wheel	https://www.calm.com/
Legal and regulatory	Equality Act 2020 (UK). Information for the public, businesses, and voluntary sectors	www.gov.uk/government/publications/equality-act-guidance
	Equality advisory service (UK). Discrimination helpline	www.equalityadvisoryservice.com
	Details of how Public Sector organisations fulfil Public Sector Equality Duty in England, Scotland, Wales and N. Ireland including policies	https://www.ucu.org.uk/media/12406/Neurodiversity-Guidance/pdf/Neurodiversity_A4_guide_January_22.pdf

(*Continued*)

Area	*Description*	*Link/details*
Access to work	The UK Government scheme for help at work for individuals who are disabled or who have a health condition	www.gov.uk/access-to-work
Psychological safety	Amy Edmondson's website, with more resources about psychological safety	https://amycedmondson.com/psychological-safety/

Inclusive technology platforms

Description	Link/details
Brain In Hand. Neurodiversity-specific assessment and assistive technology	https://braininhand.co.uk/assessors/workplace/
Braided. A platform to help make meetings more inclusive	https://braided.space/
Dragon. A speech-to-text tool	https://shop.nuance.co.uk/home
Fathom. An AI notetaker to make meetings more inclusive	https://fathom.video/

Tools

Type	Description	Link/details
Coaching style tools	**The Emotions Palette™.** Providing language to help individuals articulate their emotions	https://potent6.co.uk/product/02-emotions-palette-mini/
	Reflection cards; for different levels of conversation	https://www.holstee.com
	The Wheel of Life. A tool that you can use with your team members to assess what other support they might need in their lives	https://www.happierhuman.com/wheel-life-templates-jp2/

(Continued)

Type	Description	Link/details
Strengths-based tools	**VIA strengths and values survey**. A free assessment tool for strengths	https://www.viacharacter.org/
	Clifton StrengthsFinder. Paid strengths assessment based on research of managers	https://www.gallup.com/cliftonstrengths/en/home.aspx
	Strengths cards. To help individuals and teams to describe each other through words and pictures	https://atmybest.com/
	Strengths profiling tool. To help individuals and teams to identify their strengths and learnt behaviours	https://www.strengthsprofile.com/en-gb/about/aboutus

Index

Note: **Bold** page numbers refer to tables and *italic* page numbers refer to figures.

For Product Safety Concerns and Information please contact our EU representative GPSR@taylorandfrancis.com Taylor & Francis Verlag GmbH, Kaufingerstraße 24, 80331 München, Germany